Protection of workers
from power frequency electric
and magnetic fields

OCCUPATIONAL SAFETY AND HEALTH SERIES No. 69

PROTECTION OF WORKERS FROM POWER FREQUENCY ELECTRIC AND MAGNETIC FIELDS

A PRACTICAL GUIDE

Prepared by the International Non-Ionizing
Radiation Committee of the International Radiation
Protection Association in collaboration with the
International Labour Organization

INTERNATIONAL LABOUR OFFICE, GENEVA

Copyright © International Labour Organization 1994
First published 1994

Publications of the International Labour Office enjoy copyright under Protocol 2 of the Universal Copyright Convention. Nevertheless, short excerpts from them may be reproduced without authorization, on condition that the source is indicated. For rights of reproduction or translation, application should be made to the Publications Branch (Rights and Permissions), International Labour Office, CH-1211 Geneva 22, Switzerland. The International Labour Office welcomes such applications.

ILO
International Radiation Protection Association/International Non-Ionizing Radiation Committee
Protection of workers from power frequency electric and magnetic fields: A practical guide
Geneva, International Labour Office, 1993 (Occupational Safety and Health Series, No. 69)
/PIACT pub/, /Guide/, /Occupational safety/, /Electricity/, /Hazard/, /Radiation protection/.
13.04.2
ISBN 92-2-108261-X
ISSN 0078-3129

ILO Cataloguing in Publication Data

The designations employed in ILO publications, which are in conformity with United Nations practice, and the presentation of material therein do not imply the expression of any opinion whatsoever on the part of the International Labour Office concerning the legal status of any country, area or territory or of its authorities, or concerning the delimitation of its frontiers.
The responsibility for opinions expressed in signed articles, studies and other contributions rests solely with their authors, and publication does not constitute an endorsement by the International Labour Office of the opinions expressed in them.
Reference to names of firms and commercial products and processes does not imply their endorsement by the International Labour Office, and any failure to mention a particular firm, commercial product or process is not a sign of disapproval.

ILO publications can be obtained through major booksellers or ILO local offices in many countries, or direct from ILO Publications, International Labour Office, CH-1211 Geneva 22, Switzerland. A catalogue or list of new publications will be sent free of charge from the above address.

Printed in Switzerland   ATA

Preface

This publication is one of a series of practical guides on occupational hazards arising from non-ionizing radiation (NIR), carried out in collaboration with the International Non-Ionizing Radiation Committee (INIRC) of the International Radiation Protection Association (IRPA)[1] as part of the ILO International Programme for the Improvement of Working Conditions and Environment (PIACT).

The purpose of this book is to provide information on the possible effects of electric and magnetic fields at 50 and 60 Hz on human health and to give guidance on working conditions and procedures that will lead to higher standards of safety for all personnel engaged in the maintenance and operation of power frequency sources. It is intended for the use of competent authorities, employers and workers, and in general all persons in charge of occupational safety and health. The following topics are covered in the guide: physical characteristics of electric and magnetic fields; measurement and levels of exposure at the workplace; mechanisms of interaction; laboratory studies and observations on working populations; occupational exposure limits; and prevention and exposure control measures.

The manuscript was prepared by an IRPA-INIRC working group chaired by Professor J. Bernhardt and including Professor M. Grandolfo, Professor J. Stolwijk and Mrs. A. Duchêne. Following comments received from INIRC members, it was reviewed in detail during the annual meeting of the IRPA/INIRC in Rome, May 1991, in cooperation with Dr G. H. Coppée representing the International Labour Office.

This book is the result of a joint ILO/IRPA-INIRC activity and is published by the ILO on behalf of the two organizations. The ILO wishes to thank the IRPA-INIRC, and in particular Professor J. Bernhardt and his working group, for their contribution and cooperation in the preparation of this practical guide on the protection of workers from power frequency electric and magnetic fields.

---

[1] Since May 1992 the INIRC of the IRPA has become an independent scientific body called the International Commission on Non-Ionizing Radiation Protection (ICNIRP) and has responsibility for NIR protection in the same way as the International Commission on Radiological Protection (ICRP) has for ionizing radiation. (ICNIRP Secretariat: c/o Dipl.-Ing. R. Matthes, Bundesamt für Strahlenschutz, Institut für Strahlenhygiene, Ingolstädter Landstrasse 1, D-85764 Oberschleissheim, Germany, Tel.: +49 89 31603237, Fax +49 89 31603111.)

Contents

Preface		v
Abbreviations		x
1.	**Introduction**	1
2.	**Sources and exposure at work**	2

2.1 Sources 2
    2.1.1 General considerations 2
    2.1.2 High-voltage transmission lines 2
    2.1.3 Specific electrical equipment 3
    2.1.4 Common use of electricity 4
2.2 Electric and magnetic fields near transmission lines 4
    2.2.1 Electric fields 4
    2.2.2 Magnetic fields 10

3.   **Characteristics, measurements and physical surveillance of power frequency fields**   **14**

3.1 Physical properties 14
    3.1.1 Electric field 14
    3.1.2 Magnetic field 15
3.2 Quantities and units 16
3.3. Measurements 17
    3.3.1 Electric field-strength meters 17
    3.3.2 Calibration of electric field-strength meters 19
    3.3.3 Magnetic field-strength meters 22
    3.3.4 Calibration of magnetic field meters 22
    3.3.5 Personal dosimeters 24

4.   **Effects of power frequency electric and magnetic fields**   **25**

4.1 Electric fields: Distinction between direct and indirect interaction 25
4.2 Mechanisms of direct effects of electric fields on the human body 26
    4.2.1 Surface effects of electric fields 26
    4.2.2 Induced electric fields and currents inside the body 27
4.3 Indirect interactions of electric fields 30
    4.3.1 Short-circuit currents 30
    4.3.2 Reactions to spark discharges 34

vii

4.4	Mechanisms of field interaction of magnetic fields	35
	4.4.1 Basic mechanisms of interaction	35
	4.4.2 Induced electric fields and current densities	36
4.5	Electromagnetic interference (EMI) with cardiac pacemakers	37

5.   Human health effects of power frequency fields   **39**

5.1	Introduction	39
5.2	Studies of acute effects of ELF fields	39
	5.2.1 Observational studies of acute effects	39
	5.2.2 Studies of volunteers under controlled conditions	41
5.3	Occupational epidemiological studies of delayed effects	41

6.   Occupational exposure limits   **43**

6.1	The basis for setting protection standards	43
6.2	International recommendations	44

7.   Control of, and protection from, exposure to 50 and 60 Hz electric and magnetic fields   **47**

7.1	General considerations and scope	47
7.2	Role of the competent authorities	48
7.3	Responsibility of the employer	49
7.4	Duties of workers	50
7.5	Responsibilities of manufacturers, vendors and suppliers	50
7.6	Surveillance and monitoring of the workplace	50
	7.6.1 Survey procedures and data collection	51
	7.6.2 Assessment and interpretation of occupational exposure data	51
7.7	Control of occupational exposure	52
	7.7.1 Controlled areas	52
	7.7.2 Protective measures for electric field coupling	53
	7.7.3 Protective measures for magnetic field coupling	53
	7.7.4 Siting and installation	54
	7.7.5 Warning signs and labelling	55

Appendix A – Biological effects of ELF electric and magnetic fields   **57**

A1.	Biological effects of electric fields	58
	A1.1 Introduction	58
	A1.2 Animal experiments	58
	A1.3 *In vitro* studies	60
A2.	Biological effects of magnetic fields	60
	A.2.1 Possible field interactions	60
	A2.2 Selected *in vitro* studies	61
	A2.3 Nerve and muscle cells, nervous and cardiovascular systems	61
	A2.4 Genetic effects, reproduction and development	61

Appendix B – National exposure standards   **62**

Appendix C – Glossary   **69**

Bibliography   **73**

viii

Figures

1.	Electric field between HV-conductor and ground	5
2.	Typical overhead line and distribution of the electric field under the transmission line	6
3.	Vertically polarized electric field strength under a 765 kV, 60 Hz electric power transmission line	8
4.	Electric field strength E_0 for different types of transmission tower	9
5.	Profile of the magnetic field components under a 765 kV, 60 Hz electric power transmission line (height 15.5 m)	11
6.	Typical overhead line (380 kV) and distribution of the magnetic flux density above ground under the transmission line	12
7.	Profile of the calculated magnetic flux density (50 Hz) at 1 m above ground between the transmission towers for different phase relationships between the conductors (1 kA per phase)	13
8.	Typical shapes of E-field meter probes of the free-body type	18
9.	Typical shape of a ground-reference type E-field meter	19
10.	Scheme of the operational characteristics of electro-optic E-field meters	20
11.	Current injection calibration check	21
12.	Conducting loop in power-frequency B-field	23
13.	Distortion of the electric field by a standing human (50 Hz)	27
14.	Vibration of hairs in an alternating electric field	28
15.	Grounded human, pig and rat exposed to a vertical 60 Hz, 10 kV/m electric field	29

Tables

1.	Typical exposure levels from 50/60 Hz magnetic field sources	3
2.	Electric field strengths at heights of 0.5 and 1.5 m above ground level and at various lateral distances from typical three-phase transmission lines	7
3.	Typical exposure levels from 50/60 Hz electric field sources	10
4.	Electric and magnetic field quantities and units in the SI system	16
5.	Conversion factors for units used in practice for magnetic fields	17
6.	Induced current density ranges at 50 or 60 Hz for producing biological effects	31
7.	Effects of currents passing through the human body	33
8.	Electric field strength for person-to-vehicle currents and short-circuit body current for persons exposed to electric fields with feet grounded	34
9.	Direct and indirect effects of 50 and 60 Hz electric fields	35
10.	Limits of occupational exposure to 50/60 Hz electric and magnetic fields	45
B1.	Operation voltage and transmission route lengths of networks in countries participating in the CIGRE survey	63
B2.	Typical values of maximum electric field and Right-of-Way widths in the United States	66
B3.	Typical values of electric field strength under power lines recommended in the United States	67
B4.	Maximum permissible levels of 50 Hz magnetic fields (A/m) in the USSR (1985)	68

Abbreviations

AC	Alternating current
BEMS	Bioelectromagnetics Society
BMA	British Medical Association
cgs	centimetre, gram, second
CIGRE	International Conference on Large High-Voltage Electric Systems
ECG	Electrocardiogram
EEG	Electroencephalogram
ELF	Extremely low frequency
EMI	Electromagnetic interference
EPRI	Electric Power Research Institute (Palo Alto)
HV	High voltage
IEC	International Electrotechnical Commission
IEEE	Institute of Electrical and Electronic Engineers
ICNIRP	International Commission on Non-Ionizing Radiation Protection
INIRC	International Non-Ionizing Radiation Committee
IRPA	International Radiation Protection Association
NESC	National Electric Safety Code
NRPB	National Radiological Protection Board (United Kingdom)
PIACT	Programme for the Improvement of Working Conditions and Environment (ILO)
RoW	Right-of-Way
rms	root-mean-square
UNEP	United Nations Environment Programme
UNIPEDE	International Union of Producers and Distributors of Electrical Energy
WHO	World Health Organization

1

Introduction

Just over 100 years ago, human exposure to external electric and magnetic fields was limited to those fields arising naturally from both extraterrestrial and terrestrial sources. Within the past 50 years, there has been very significant growth of manmade extremely low frequency (ELF) electric and magnetic fields at frequencies of 50 and 60 Hz (power frequencies), predominantly from electric power generating and distribution systems. Manmade ELF fields are now greater than the natural fields by many orders of magnitude.

Within all organisms are endogenous electric fields and currents that play a role in the complex mechanisms of physiological control, such as neuromuscular activity, glandular secretion, cell membrane function, and tissue growth and repair. It is not surprising that, because of the role of electric fields and currents in so many basic physiological processes, questions arise concerning possible effects of these artificially produced fields on biological systems. Indeed, with the ever-increasing need for power transmission, workers' concern about these fields continues to grow, especially since scientists seem unable to provide unreserved assurance of safety.

This guide comprises a review of data on the effects of ELF electric and magnetic fields at 50 and 60 Hz on biological systems pertinent to the evaluation of health risks for workers. One of its purposes is to provide information on the possible effects of exposure to 50/60 Hz electric and magnetic fields on human health, and to give guidance on the assessment of risks from occupational exposures. Areas in which uncertainties exist and further research is needed are also indicated.

As the guide mainly concerns effects directly attributed to electric and magnetic fields, it discusses only briefly the effects of co-generated ozone, noise, ultraviolet radiation and X-rays from corona discharges, and induced short-circuit currents which may be important factors in the overall environment. In general, the effects of contact currents have not been considered in detail since restriction of leakage currents is already dealt with in a number of standards.

2

Sources and exposure at work

2.1 Sources

2.1.1 General considerations

With the advent of the technological age, the widespread use of electrical appliances, electromagnetic energy and high-voltage wires has markedly increased environmental exposures to 50/60 Hz electric and magnetic fields.

The principal manmade sources of 50/60 Hz electric and magnetic fields are high-voltage transmission lines, and all devices containing current-carrying wires, including equipment and appliances in industry and in the home, operating at power frequencies of 50 Hz in most countries and 60 Hz in North America.

The natural background electric field strength at the power frequencies of 50 or 60 Hz is about 10^{-4} V/m, which means that fields in the close vicinity of high-voltage (HV) transmission lines are 10^8 times stronger, and fields introduced into homes by wiring or appliances are still about 10^3 to 10^6 times stronger than the natural background.

2.1.2 High-voltage transmission lines

The electric fields associated with high-voltage transmission lines, sub-stations and certain industrial equipment are the strongest electric fields to which persons, animals and plants may be routinely exposed.

High-voltage lines are operated at standard voltages such as 110, 138, 220, 230, 345, 380, 500, 735 or 750 kV, or 1,150 kV in the USSR[1] (50 Hz), but the construction of 1-1.2 MV, or even up to 1.5 MV lines is in progress or at various stages of planning.

Magnetic fields associated with overhead transmission lines are usually relatively low and have so far attracted less attention. Stronger magnetic fields occur near large magnets, electrical furnaces or other devices using high currents. The potential use of cryogenic or superconducting transmission facilities would introduce another localized source of strong magnetic fields.

---

[1] The names of the countries used in this text are consistent with the dates of the relevant texts and standards cited.

SOURCES AND EXPOSURE AT WORK

Utility employees working in substations or on maintenance of transmission lines form a special group continually exposed to large fields.

Environmental exposures are often categorized as acute or chronic. People in their homes are chronically exposed to electric fields of at least a few volts per metre, and people spending extended periods of time near high-voltage equipment or working for power companies maintaining such equipment may be acutely exposed for several days a week to fields of several thousand volts per metre. Similarly, most of the population are chronically exposed to magnetic flux densities of less than 1 μT. People working near high-current machines are sometimes exposed at levels exceeding 0.1 mT over a substantial part of the work-day.

2.1.3 Specific electrical equipment

Small groups of workers in the electrochemical industry and in research laboratories may be exposed to high electric and magnetic field strengths. For instance, near induction furnaces and industrial electrolytic cells magnetic flux densities can be measured as high as 50 mT.

Occupational exposure to magnetic fields stems predominantly from working near industrial equipment using high currents. Such devices include various types of welding machines, electroslag refining, various furnaces, induction heaters and stirrers. Details of surveys of magnetic field strengths in industrial settings are given in table 1. Information on average and maximum values is given by taking note of the fact that magnetic fields are often very inhomogeneous. Surveys on induction heaters used in industry show magnetic flux densities at operator locations ranging from 0.0007 to 6 mT, depending on the frequency used and the distance from the machine. In their study of magnetic fields from industrial electrosteel and welding equipment, Lövsund et al. (1982) found that spot welding machines and ladle furnaces produced fields up to 10 mT at distances up to 1 m.

Table 1. Typical exposure levels from 50/60 Hz magnetic field sources

Magnetic field source	Magnetic flux density (mT)	
	Average level	Peak level
High-voltage AC power transmission:		
Overhead transmission lines	0.01–0.03	0.4
(380 and 765 kV)		(during failure)
Generating stations	0.02–0.04	0.27
Industry and medicine:		
Electrolytic processes	1–10	50
Welding machines	0.1–5.8	130
Induction heating	1–6	25
Ladle furnace	0.2–8.0	–
Arc furnace	up to 1	–
Electroslag welding	0.5–1.7	–
Therapeutic equipment	1–16	–
Office and household:		
Standard dwellings	10^{-5}–10^{-3}	10^{-3}–4×10^{-2}
Residences with electric heating	1.2×10^{-2}	–
	(at 30 cm)	

Sources: Lövsund et al., 1982; Krause, 1986; Stuchly, 1986; and Tenforde, 1986.

2.1.4 Common use of electricity

The use of electricity (50/60 Hz and 110/220-240 V) is widespread all over the world in both industrialized and developing countries, and in all sectors of activity. It is an integral part of the life and work of millions of people.

The use of electricity includes the hazards of electrocution and fire, which must be prevented. Substantial knowledge has been accumulated and a multitude of laws, regulations and standards have been enacted and published over time in all countries to protect workers and the public. They concern protection against accidents and are designed to ensure safety in the use of electricity.

Progress in medical knowledge and research has shown that high-strength power frequency electric and magnetic fields cause biological effects. This raises two basic questions:

(a) Are these biological effects significant from a health point of view, i.e. will they lead to harmful health effects, and to health impairment and disease?

(b) Is the knowledge derived from experience with high-strength 50/60 Hz electric and magnetic fields applicable to exposures to fields of lower strength?

Current scientific knowledge suggests that there are no detectable health effects and that there is no reason to worry about health impairment due to exposure to electric and magnetic fields associated with the common use of electricity in the household, offices and industry. The problem of present concern is the prevention of electrical accidents rather than hypothetical risks due to exposure to very low electric and magnetic field strengths.

2.2 Electric and magnetic fields near transmission lines

2.2.1 Electric fields

For 50/60 Hz, the electric and magnetic fields must be considered separately. For electric fields produced by power systems, it is useful to distinguish the fields by their ability to cause current to flow in nearby objects (Miller, 1974). High-impedance electric fields are those which couple to objects capacitively and, thus, induce weak currents in them. Low-impedance fields in a conductive medium, such as the earth, can cause currents to flow in objects which contact the earth.

Power lines have high-impedance fields normal to the axis of the line. The electric field lines begin on the conductors and fringe to other nearby conductors which are at different potentials. These electric fields fringe to the earth, the metal towers supporting the wires, the other wires, nearby railway tracks, and so on. While this field may be intense under high-voltage transmission lines, it can induce only relatively weak currents in nearby objects. Under any power line, the total electric field is the superposition of the contributions of each conductor.

SOURCES AND EXPOSURE AT WORK

Current flow in a long conductor near the surface of the earth induces a longitudinal electric field in the earth. Again, the total longitudinal field under a power line is the superposition of the components induced by the currents in each wire.

Finally, there is an electric field which appears in the vicinity of power system ground terminals. This field is caused by the flow of current into the earth at a substation, pole or home ground.

Electrical energy is transmitted from the power plant, where it is generated, along conductive, metallic transmission connections (overhead power lines or underground cables), to substations and finally to energy consumers. The electric field lines (the directions along which a charge is moved by the force imposed by the field) between an HV-conductor and ground are shown in figure 1.

Figure 1. Electric field between HV-conductor and ground

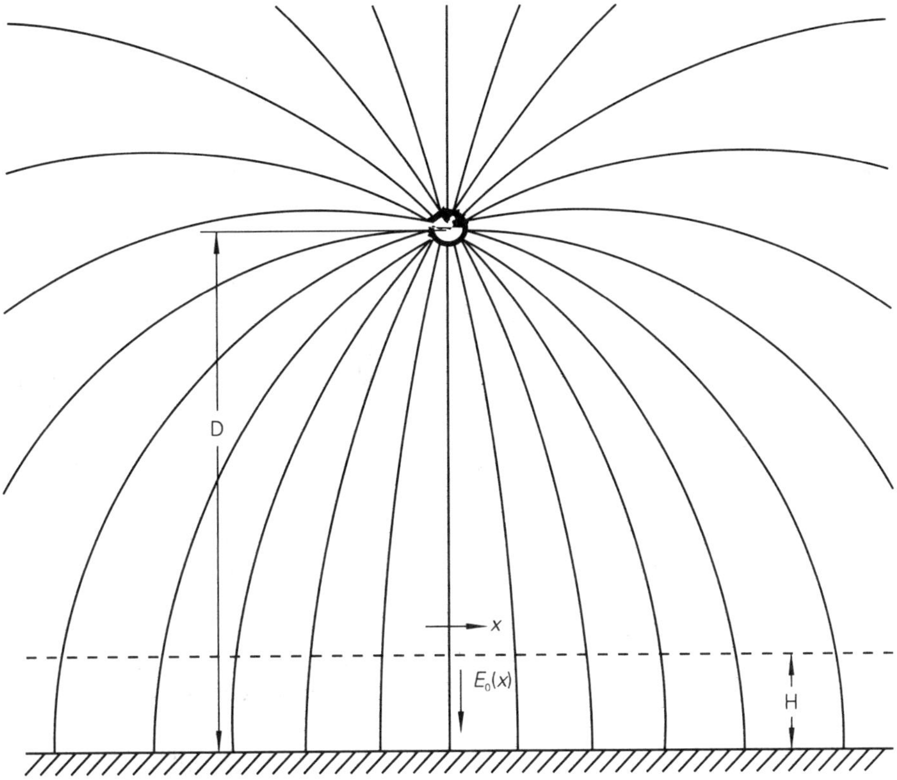

Source: K.-H. Schneider: "Elektrische und magnetische Felder", in *Strahlenschutz in Forschung und Praxis* (Stuttgart, Georg Thieme Verlag), No. 20, 1980, pp. 30-33.

Figure 2. Typical overhead line and distribution of the electric field under the transmission line

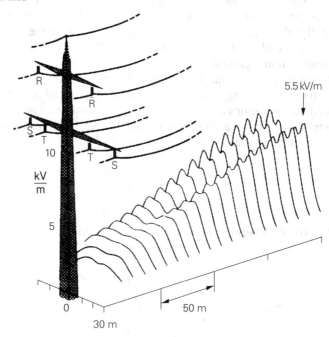

Source: H.-J. Haubrich: "Biologische Wirkungen elektromagnetischer 50 Hz-Felder auf den Menschen", in *Elektrizitätswirtschaft*, Vol. 86, Nos. 16/17, 1987, pp. 697-705.

A typical overhead line (figure 2) consists of structures (transmission towers or pylons) from which the live conductors are suspended by sets of insulators. The conductors of each phase are suspended far enough away from the other conductors and the transmission tower to prevent flashover or short-circuiting between one phase and another, or between the phases and earth (via the supporting structure). In overhead lines, the conductors consist of bare metal cables. Thus, any approach to a live conductor presents a lethal danger due to flashover and a resulting electric current flow that would precede actual contact with a conductor.

High-voltage lines are operated at standard voltages up to 750, 765 or 1,150 kV. The construction of higher-voltage lines up to 2,000 kV is in progress or at various stages of planning.

Alternating current (AC) three-phase HV lines are most widely used. One circuit of the three-phase line comprises three single or three sets of conductors under high voltage and one or three grounded conductors that protect the live conductors against lightning.

At ground level, beneath high-voltage transmission lines, the electric fields created have the same frequencies as those carried by the power lines. The characteristics of these fields depend on the line voltage, and on the geometrical dimensions and positions of the conductors of the transmission line. The field intensity selected for reference or comparison purposes is the unperturbed ground-level electric field strength. To avoid the effects of vegetation or irregularities in the terrain, the unperturbed field strength is usually computed or measured at a given height above ground level (0.5, 1, 1.5 or 1.8 m).

There are several primary influences on the electric field strength beneath an overhead transmission line. These include:

(a) the height of the conductors above ground;

(b) the geometric configuration of conductors and earthing wires on the towers, and in the case of two circuits in proximity, the relative phase sequencing;

(c) the proximity of the grounded metallic structure of the tower;

(d) the proximity of other tall objects (trees, fences, etc.);

(e) the lateral distance from the centre line of the transmission line;

(f) the height above ground of the point of measurement;

(g) the actual (rather than the nominal) voltage on the line.

Conductor height, geometric configuration, lateral distance from the line, and the voltage of the transmission line are by far the most significant factors in considering the maximum electric field strength at ground level. At lateral distances of about twice the line height, electric field strength decreases with distance in an approximately linear fashion (Zaffanella and Deno, 1978).

Tell (1983) reported results of extensive calculations and measurements of electric field strengths near ground level under extra-high-voltage electric power lines.

Table 2. Electric field strengths at heights of 0.5 and 1.5 m above ground level and at various lateral distances from typical three-phase transmission lines

Line voltage (kV)	Lateral distance from line (m)	E-field at 0.5 m above ground level (V/m)	E-field at 1.5 m above ground level (V/m)
110	0	1 560	–
	10	540	–
	20	150	–
	30	50	–
750–765	0	9 000	21 000
	10	8 000	9 000
	20	2 500	3 500
	30	1 000	1 800
	50	300	350
1 300	0	15 000	29 000
	10	11 000	15 000
	20	3 000	6 000
	30	1 000	2 500

Figure 3 illustrates the vertically polarized electric field strength under a 765 kV, 60 Hz transmission line. Both calculated and measured data show good agreement. Nominal maximum values of electric field of about 10 kV/m can exist beneath 765 kV lines at 1 m above ground.

The electric field strengths at 0.5 m and 1.5 m above ground level at various lateral distances from typical three-phase transmission lines are given in table 2.

The electric field strengths at 0.5 m above ground level from various transmission lines are shown in figure 4.

Inside buildings near HV transmission lines, the field strengths are typically lower than the unperturbed field by a factor of about 10-100, depending on the structure of the building and the type of materials.

Occupational exposures that occur near high-voltage transmission lines depend on the worker's location, either on the ground or at the conductor during live-line

Figure 3. Vertically polarized electric field strength under a 765 kV, 60 Hz electric power transmission line

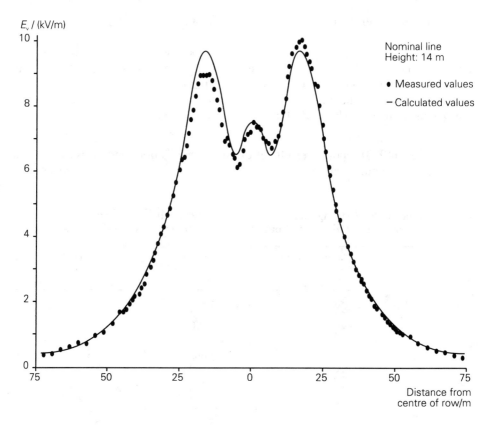

Source: M. Grandolfo et al.: *Biological effects and dosimetry of static and ELF electromagnetic fields* (New York and London, Plenum Press, 1985), p. 62, fig. 7.

Figure 4. Electric field strength E_o for different types of transmission tower

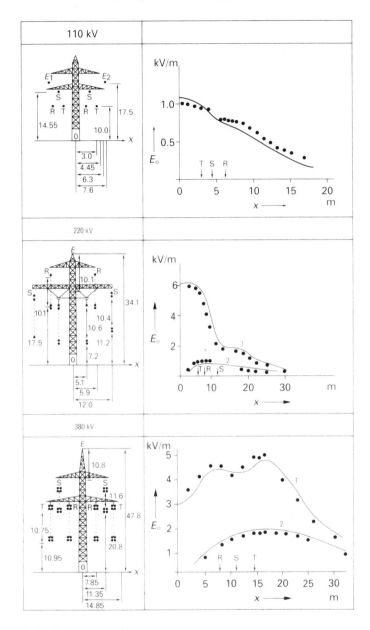

Note: ——— = calculated; ... = measured.

Values 0.5 m above ground; distance x from axis of line (m); RST conductor phases; E earthed conductor; 1 = values between the transmission towers; 2 = values near a transmission tower.

Source: K.-H. Schneider et al.: *Displacement currents to the human body caused by the dielectric field under overhead lines*, CIGRE Report 36-07 (Paris, Conférence Internationale des Grands Réseaux Electriques, 1974); reproduced by kind permission of CIGRE.

work at high potential. Typical exposure levels in HV-generating stations are shown in table 3. When working under live-line conditions, protective clothing may be used to reduce the electric field strength and current density in the body to values similar to those that would occur for work on the ground. Protective clothing does not weaken the influence of the magnetic field.

A high-voltage conductor can create ozone (O_3) by means of ionization of air near its surface. Under certain weather conditions, causing corona discharges in the vicinity of HV transmission lines, the formation of ozone occurs. However, since ozone is a very unstable gas, it rapidly decomposes into harmless oxygen compounds in the open air and biological effects should not be expected. Measurement and calculations of ozone near transmission lines show that local increments in ozone levels are insignificant.

Table 3. Typical exposure levels from 50/60 Hz electric field sources

Electric field source (kV)		Electric field strength
Office and household levels:		
Standard dwellings		2–500 V/m (at 30 cm)
High-voltage AC power transmission:		
Overhead transmission		
lines	110	1–2 kV/m
	245	2–3 kV/m
	380	5–6 kV/m
	800	10–12 kV/m
Generating stations		
	110	5–6 kV/m
	245	9–10 kV/m
	380	14–16 kV/m
	800	14–16 kV/m
Source: Grandolfo et al., 1985.		

Noise is of concern beneath or near power lines, and in switchyards. Techniques are available to reduce corona-induced noise beneath power lines to acceptable levels or to standards set by law. Both the frequency spectrum of the noise and the intensity in different spectral regions should be taken into account. Effects of noise in terms of annoyance may create problems, particularly in densely populated urban and suburban areas. In switchyards the acoustic environment is special and can differ considerably from that near the overhead lines; in addition, ozone, nitrogen oxides and broadband electromagnetic fields, due to corona discharges, may be present.

2.2.2 Magnetic fields

In the absence of ferromagnetic materials, the magnetic field lines are solenoidal, i.e. they form concentric circles around the conductor.

SOURCES AND EXPOSURE AT WORK

Apart from the geometry of the conductor, the maximum magnetic flux density is determined only by the magnitude of the current. The magnetic field beneath high-voltage overhead transmission lines is directed mainly transverse to the line axis. The maximum flux density at ground level may be under the centre line or under the outer conductors, depending on the phase relationship between the conductors. The maximum magnetic flux density at ground level for a double-circuit 500 kV overhead transmission lines system is approximately 35 µT per kiloampere. The field at ground level beneath a 765 kV, 60 Hz power line carrying 1 kA per phase is 15 µT. Figure 5 shows the profile of the three magnetic field components, measured at a height of approximately 1.5 m above the ground, under a 765 kV, 60 Hz line characterized by a height of approximately 15.5 m.

Typical values for the magnetic flux at worksites near overhead lines, in substations and in power stations (16 2/3, 50, 60 Hz) range up to 50 µT (Krause, 1986).

Figure 5. Profile of the magnetic field components under a 765 kV, 60 Hz electric power transmission line (height 15.5 m)

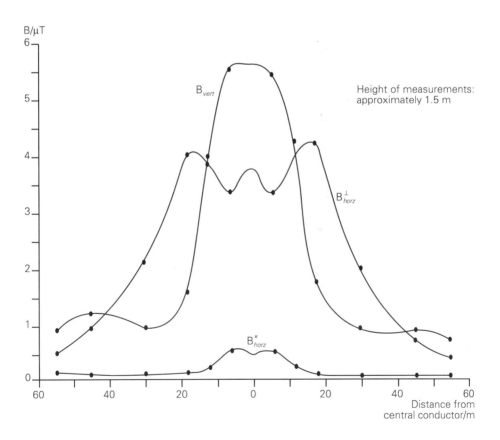

Source: Grandolfo et al., 1985, p. 64, fig. 8.

11

Figure 6 shows a typical 380 kV overhead line and the distribution of the magnetic flux density above ground under the transmission line. Figure 7 shows a profile of the calculated magnetic flux density (50 Hz) at 1 m above ground between the transmission towers for different phase relationships between the conductors (1 kA per phase). The magnetic flux density decreases with distance from the conductor to values of the order of 1 to 10 μT at a lateral distance of about 20-60 m from the line (see figure 7).

Figure 6. Typical overhead line (380 kV) and distribution of the magnetic flux density above ground under the transmission line

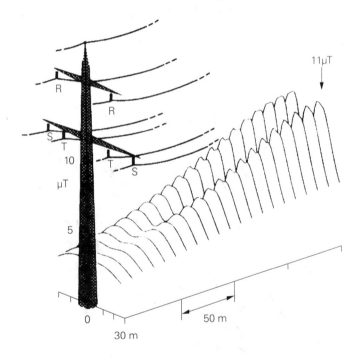

Source: As for figure 2.

Figure 7. Profile of the calculated magnetic flux density (50 Hz) at 1 m above ground between the transmission towers for different phase relationships between the conductors (1 kA per phase)

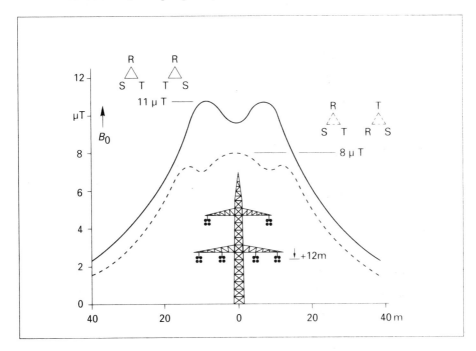

Source: As for figure 2.

3

Characteristics, measurements and physical surveillance of power frequency fields

3.1 Physical properties

It is useful to review briefly some basic concepts of fundamental physics which are essential for developing an analysis of the characteristics of electromagnetic radiation, and in particular of 50/60 Hz electric and magnetic fields. The experimental evidence of electric and magnetic phenomena is consistent with the following postulates:

1. There exist two kinds of electric charge: we call them positive and negative.

2. Electric charge is conserved. This means that in any isolated system the total electric charge is constant: whenever a positive charge appears or disappears, an equal amount of negative charge must also appear or disappear.

3. An electric charge in motion undergoes a force consisting of two components, either of which, or both, may eventually be zero: the first one is independent of its speed and is proportional to a quantity E which is termed electric field; the second one is proportional to its speed and to a quantity H which is termed magnetic field.

3.1.1 Electric field

Electric charges exert forces on each other. It is convenient to introduce the concept of an electric field to describe this interaction. Thus, we say that a system of electric charges produces an electric field at all points in space (this field propagates away from the source charges at the speed of light) and that any other charge placed in the field will experience a force because of its presence. The electric field is denoted E and is a vector quantity, which means that it has both strength and direction. The force, F, exerted on a point body containing an electric charge q placed in an electric field E, is given by q times the electric field strength.

The force on a positive charge (e.g. a proton) is in the same direction as E, while the force on a negative charge (e.g. an electron) is in the opposite direction. The unit of electric charge is the coulomb (C). The smallest quantity of electric charge that has been observed in nature is that of an electron or proton, $1.6 \cdot 10^{-19}$C. It appears that all larger quantities of charge are integral multiples of the electronic charge. The unit of E-field strength is newton/coulomb.

CHARACTERISTICS, MEASUREMENTS AND PHYSICAL SURVEILLANCE

It is generally easier and more useful to measure the electric potential, V, rather than the E-field. This is because the potential is much less dependent on the physical geometry of a given system (e.g. location and sizes of conductors). Therefore, in practice, for the electric field strength the unit volt/metre (V/m) is used.

Electric fields exert forces on charged particles. In an electrically conductive material, such as living tissue, these forces will set charges into motion to form an electric current. This current is frequently specified by the current density vector, J, whose magnitude is equal to the current flowing through a unit area oriented perpendicular to its direction.

The unit of current density, J, is A/m^2. J is directly proportional to E in a wide variety of materials, where the constant of proportionality is called the electrical conductivity of the medium. The unit of conductivity is siemens/metre (S/m).

3.1.2 Magnetic field

The fundamental vector quantities describing a magnetic field are the field strength, H, and the magnetic flux density, B (or equivalently, the magnetic induction).

Magnetic fields, like E-fields, are produced by electric charges, but only when they are in physical motion. Magnetic fields exert forces on other charges but, again, only on charges which are in physical motion. Since the most common manifestation of electric charge is motion in an electric current, it is often said that H- and B-fields are produced by electric currents and interact with other electric currents.

The force F acting on an electric charge q moving with a velocity v perpendicular to a magnetic flux density B is given by $q \times B \times v$, where the direction of F is perpendicular to *both* v and B. If the direction of v were, instead, parallel to B, then F would be zero in this example. This illustrates one important characteristic of a magnetic field: it does no physical work, because the force, called the Lorentz force, generated by its interaction with a moving charge is always perpendicular to the direction of motion.

The magnetic field strength (H) is the force with which the field acts on an element of current situated at a particular point. The value of H is measured in ampere per metre (A/m). The trajectories of the motion of an element of current (or the orientations of an elementary magnet) in a magnetic field are called the magnetic lines of force.

As in the case of electric fields, single-phase and three-phase magnetic fields can be defined: the field at any point may be described in terms of its time-varying magnitude and invariant direction (single phase), or by the field ellipse, i.e. the magnitude and direction of the major and minor semi-axes (three phases).

The magnetic flux density, B, rather than the magnetic field strength, H, is used to describe the magnetic field generated by currents in the conductors of transmission lines and substations. The magnetic flux density B is defined as the product of H times the value of the magnetic permeability which is determined by the properties of the medium and, for most biological material, is equal to μ_0, the value of permeability of free space (air). Thus, for biological materials, the values of B and H are related by a constant (μ_0).

3.2 Quantities and units

Dealing with exposure to electric and magnetic fields, a number of physical quantities are currently used (IRPA/INIRC, 1985; UNEP/WHO/IRPA, 1984). Table 4 lists the main ones, together with the corresponding SI units and symbols.

Table 4. Electric and magnetic field quantities and units in the SI system

Quantity	Symbol	Unit
Frequency	f	hertz (Hz)
Electric field strength	E	volt per metre (V/m)
Electric flux density	D	coulomb per square metre (C/m²)
Capacitance	C	farad (F)
Current	I	ampere (A)
Current density	J	ampere per square metre (A/m²)
Electric charge	Q	coulomb (C = A · s)
Impedance	Z	ohm (Ω)
Volume charge density	ρ	coulomb per cubic metre (C/m³)
Magnetic field strength	H	ampere per metre (A/m)
Magnetic flux	ϕ	weber (Wb) = V · s
Magnetic flux density	B	tesla (T) = Wb/m²
Permittivity	ϵ	farad per metre (F/m)
Permittivity of vacuum	ϵ_0	$\epsilon_0 = 8.854 \cdot 10^{-12}$ F/m
Permeability	μ	henry per metre (H/m)
Permeability of vacuum	μ_0	$\mu_0 = 1.257 \cdot 10^{-6}$ H/m
Time	t	second (s)

Although they may be familiar to most people, a few remarks about the units may be useful. Units of the *Système international d'unités* (SI) have become conventional, and in many circumstances compulsory, for expressing physical quantities. SI is based on seven independent quantities: length (m), mass (kg), time (s), electric current (A), thermodynamic temperature (K), luminous intensity (cd), and amount of substance (mol). Therefore among the electric units the coulomb is a derived unit expressed in terms of ampere and second, and not vice versa. In general, electric units are well known and currently used. That is not the case for magnetic units, because many researchers are still accustomed to some centimetre, gram, second (cgs) units. It is therefore to be recalled, for example, that the magnetic flux density, which is accepted as the most relevant quantity to describe magnetic fields, is measured in tesla (T). The tesla is defined in terms of the force between parallel wires carrying electric current: thus 1 tesla is equal to 1 newton per ampere and per metre. A field having a magnetic flux density, B, of 1 T in SI units would have a value of 10^4 gauss in the cgs system. Sometimes it may be preferable to specify the magnetic field strength, H, expressed in ampere per metre. A field strength of 1 A/m is equivalent to $4\pi \cdot 10^{-3}$ oersted in the cgs system. It is important to note that a physically relevant distinction between B and H becomes apparent only in a medium which has a net polarization of

CHARACTERISTICS, MEASUREMENTS AND PHYSICAL SURVEILLANCE

magnetic dipoles. In free space, and within a very good approximation also in air and in any biological system, B and H are proportional, where the constant of proportionality is $\mu_0 = 4\pi \cdot 10^{-7}$ henry per metre (H/m). For any practical purpose we can speak in terms of tesla or in terms of ampere per metre, recalling that $1\ T = 10^7/4\pi$ A/m. Finally, the magnetic flux within a given surface is the product of the area and the component of magnetic flux density normal to its surface. The corresponding SI unit is the weber; 1 Wb equals $1\ T \cdot m^2$.

For convenience, the conversion factors relating the various units used in practice for magnetic fields are given in table 5.

Table 5. Conversion factors for units used in practice for magnetic fields

From/To	T*	G	γ	A/m	Oe
T*	1	10^4	10^9	$7.96 \cdot 10^5$	10^4
G	10^{-4}	1	10^5	79.6	1
γ	10^{-9}	10^{-5}	1	$7.96 \cdot 10^{-4}$	10^{-5}
A/m	$1.257 \cdot 10^{-6}$	$1.257 \cdot 10^{-2}$	1 257	1	$1.257 \cdot 10^{-2}$
Oe	10^{-4}	1	10^5	79.6	1

Symbols : G = gauss; γ = gamma; A = ampere; m = metre; Oe = oersted; T = tesla; Wb/m² = weber per square metre.

3.3 Measurements

3.3.1 Electric field strength meters

The principal types of instrumentation for the measurement of the electric field strength are the following (Conti, 1985):

(a) the self-contained, or free-body type;

(b) the ground-reference type; and

(c) the electro-optic type.

An instrument type is characterized by its operational characteristics. Basically, an electric field strength meter has two parts; the probe or field sensor, and the detector which consists of signal-processing circuitry and an analogue or digital display.

17

Electric field meters are calibrated in an almost uniform electric field so that, in principle, their use is not correct for the measurement of very non-uniform fields, such as those that are perturbed by the proximity of sharply curved objects. For the same reason, when performing measurements, particular care should be taken to avoid or, at least, to limit the proximity effect due to the operator.

Electric field meters are calibrated to read the root-mean-square (rms) value of the electric field strength along the vertical axis of the probe.

Free-body field strength meters operate by measuring the power frequency induced current or charge oscillating between two halves of an isolated conductive body introduced into the field to be measured.

For this type of instrument the detector is in general contained in, or is an integral part of, the probe. There also exist free-body meters suitably designed for remote display of the field strength measured. In this case, the detector consists of two parts: one, contained in the probe, converts the current signal induced in the probe to a proportional optic signal, while the other, connected to the first through a fibre optic link, processes the optic signal for the purposes of monitoring the electric field strength (Armanini and Brambilla, 1979).

The probe is introduced into the electric field on an insulating handle or on a dielectric support. Free-body field strength meters are battery operated.

The shapes and the above-mentioned characteristics of this type of instrument make it portable and, consequently, suitable for survey-type measurements. Typical shapes of free-body meters are shown in figure 8.

It should be noted that the electric field direction serves as an alignment axis for the probe and that during field measurements this axis should be aligned with the field component of interest.

Figure 8. Typical shapes of *E*-field meter probes of the free-body type

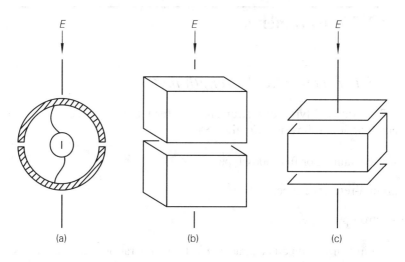

Source: Reproduced from IEC 833 (1987) – *Measurement of power-frequency electric fields*. Copyright retained by the International Electrotechnical Commission, Geneva.

The ground-reference electric field strength meter is only used to measure field strength at ground level. The probe consists of two parallel plates separated by a thin insulating sheet as shown in figure 9.

The use of this type of meter is restricted to flat ground planes. It may be battery or mains powered, but requires a ground reference potential.

The electro-optic type field meter utilizes Pockel's effect in a dielectric crystal probe (Pockel-cell) for determining the electric field strength (Hamasaki et al., 1980) – figure 10. The electric field induces an optical birefringence in a properly oriented dielectric crystal, the magnitude of which is proportional to the field strength.

Like the free-body meter, it is suitable for survey-type measurements because it is portable, allows measurements above the ground plane and does not require a ground reference potential. The dielectric probe and detector are connected with fibre optics. Light from a source in the detector is transmitted to and from the field-sensing probe through the fibre-optic connections. The probe can be introduced into the field with a dielectric rod. Because of the small dimensions of the probe (about 2 cm), the likelihood of perturbing charge distributions on boundary surfaces is small. In addition to this, the small dimensions of the probe and its dielectric characteristics make these meters particularly useful for measurement of very uniform fields.

3.3.2 Calibration of electric field-strength meters

Three methods are generally used for calibration purposes (Conti, 1985):

(a) calibration of the meters in a uniform electric field of known field strength, suitably produced by electrode systems;

Figure 9. Typical shape of a ground-reference type *E*-field meter

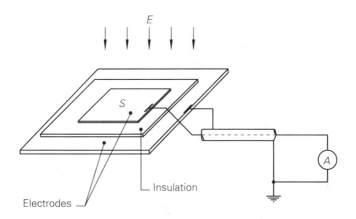

Source: As for figure 8.

Figure 10. Scheme of the operational characteristics of electro-optic E-field meters

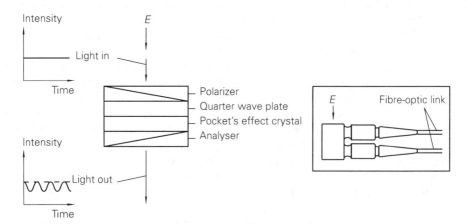

Source: As for figure 8.

(b) calibration of the meters under a transmission line, using a device quite similar to the ground-reference-type electric field-strength meter;

(c) calibration of the meter by using current-injection circuits.

The dimensions of the apparatus generating the calibration field must be sufficiently large, so that the charge distributions on the boundary surfaces are, at worst, only slightly perturbed when the probe is introduced. In addition, the volume of uniformity must be sufficiently large to reduce the uncertainty in the value of the field strength at the location of the probe to an acceptable level.

Any nearby objects possibly perturbing the field must be removed while performing the calibration, and the operator should be sufficiently far away from the apparatus.

Since field-strength meters, with the exception of the electro-optic type, are frequency dependent, the frequency of the source used for calibration should be the same as that of the field to be measured, and the power supply should be almost free (< 1 per cent) of harmonic content. In the event of such a power supply not being available, the error caused by the presence of harmonics should be quantified. Dielectric handles used during field-strength measurements should also be used to support the probe during calibration.

Transmission lines produce electric fields at ground level that may be used for calibration purposes.

Current injection, rather than a calibration method, may be considered as a method of checking calibration. Indeed it requires that the ratio of induced current to electric field strength, I/E, for a given free-body type (excluding the electro-optic type) or ground-reference type field meter, should be previously determined using, for example, a parallel plate apparatus.

A circuit such as that shown schematically in figure 11 can be used to inject a known current *I* on to the probe-sensing electrodes of the electric field-strength meter to be calibrated. *V* is a precision voltmeter and Z is a known impedance at least two orders of magnitude greater than the input impedance of the electric field-strength meter. The induced current can thus be calculated from Ohm's law to an accuracy of within 0.5 per cent. Although resistors or capacitors may be used as the impedance shown in figure 11, the use of resistors is recommended. Resistors are preferred because the admittance of capacitors increases with frequency. Therefore, the presence of harmonics in the source wave-form can lead to greater errors than if resistors are used.

A circuit similar to that shown in figure 11 may be used for checking the calibration of ground-reference type field meters. In this case the impedance of the ground side of the circuit is removed, and the remaining impedance is doubled in value.

If current injection is used, particular care should be taken to eliminate signal contribution from interfering ambient fields, possibly produced by nearby sources, such as power cables or power supplies.

Figure 11. Current injection calibration check

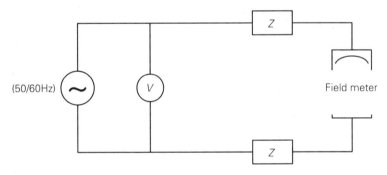

Source: Institute of Electrical and Electronics Engineers (IEEE): *IEEE standard procedures for measurement of power frequency electric and magnetic fields from AC power lines,* an American National Standard, Std. 644-1987 (New York).

Measurement uncertainty during practical outdoor measurements (e.g. under transmission lines, or substation busbars), using commercially available free-body type meters, may approach 10 per cent.

It is useful to summarize the main rules to be followed for the correct performance of measurements of power-frequency electric field strength (Conti, 1985):

– Measurements should be performed in such a way as to avoid detectable interactions between the probe and electrodes which generate the electric field.

– Observer-to-probe distance should be kept greater than 1.5 m in order to limit the observer proximity effect to less than 5 per cent.

- The height of the probe above the ground plane should be kept greater than twice its largest diagonal dimension in order to avoid significant interaction between the probe and the ground plane.

- Any objects permanently or temporarily placed at the measurement location will produce a perturbing effect. If possible, the non-permanent objects should be removed, while objects that cannot be removed should have their dimensions and location in relation to the electrodes generating the field (conductors of overhead lines, busbar system in substation) recorded.

- Handle and dielectric support should be dry and clean.

- In the case of battery-operated meters, the state of the power source should be checked before carrying out measurement.

3.3.3 Magnetic field-strength meters

Magnetic field probes consisting of electrically shielded coils of wire are used in conjunction with portable voltmeters to measure power-frequency magnetic fields. Other probes are also available that can measure the magnetic flux density (B-field) by measuring the resistance of resistors, which vary in value depending on B, or by measuring the Hall effect in semi-conducting materials (Armanini, 1970). This type of instrument is, however, less used than the first type mentioned, at least for measurements of the power-frequency magnetic field, and will not be considered here.

Unlike electric field measurements, less experience has been acquired in the measurement of B-fields under transmission lines. This relative inexperience is to some extent offset by fewer mechanisms for B-field perturbations and measurement errors as compared with electric fields. The instrumentation considered here consists of a shielded-coil probe and a shielded detector with a connecting shielded cable. The probe can be held with a short dielectric handle without seriously affecting the measurement. Proximity effects due to dielectrics and poor non-magnetic conductors are in general negligible.

The operating principle of a coil B-field probe can be explained by considering a closed loop of a conductor with area A immersed in a quasi-static, uniform magnetic flux density B, as shown in figure 12. An electromotive force is induced in the loop (and a current will flow) as a consequence of changes in magnetic flux ϕ (B) through the area A, which is directly proportional to the time rate of ϕ (B). Measurement of the induced electromotive force provides a measure of the B-field when B is assumed to be uniform and to have its direction perpendicular to the plane of the loop.

Earlier remarks regarding the response of the detector to the power frequencies and harmonic components of the E-field apply also to the present case.

3.3.4 Calibration of magnetic field meters

Calibration of magnetic field probes is usually done by introducing the probe into a nearly uniform magnetic field of known magnitude and direction (Conti, 1985).

Figure 12. Conducting loop in power-frequency B-field

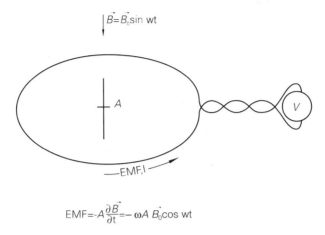

Source: Grandolfo et al., 1985, p. 206, fig. 11.

Helmholtz coils have frequently been employed to generate such fields, but a more simply constructed single loop (of many turns) with rectangular geometry has also been used. Simplicity in construction is at the expense of reduced field uniformity, but sufficient accuracy is nevertheless easily obtained.

Many of the difficulties described for making electric field measurements are not of serious concern for B-field measurements. Positioning the probe, reading errors, proximity effect of the observer or nearby (non-conducting) objects, electrical leakage of the probe handle and non-uniformity of the field have much less or negligible impact. Electrostatic shielding of the probe, however, is essential to avoid induced currents due to the ambient electric field.

Temperature effects on the detector and mechanical balance of the meter movement remain possible sources of error.

For measurements of power-frequency magnetic fields, the probe should be oriented in space for the maximum reading. The operator may stay close to the probe. Non-permanent objects containing magnetic materials or non-magnetic conductors should be at least three times the largest dimension of the object away from the point of measurement in order to measure the unperturbed field value. The distance between the probe and permanent magnetic objects should not be less than 1 m, in order to measure the ambient perturbed field accurately.

Non-magnetic metal objects will develop eddy currents because of the time variation of magnetic flux. The magnetic fields generated by these eddy currents will decrease as the inverse third power of distance for distances large compared to the dimensions of the conductor.

To provide a more complete description of the B-field at a point of interest, measurement of the maximum and minimum fields, with their orientations in the plane of the field ellipse, can be made.

3.3.5 Personal dosimeters

Recently, exposure meters have appeared on the market which are able to measure, record and analyse power-frequency electric and magnetic fields. These personal dosimeters are a powerful tool which can be used to facilitate the collection of data, especially in epidemiological studies.

Personal dosimeters are pocket-sized, battery-operated electronic instruments, and are designed to monitor immediate personal environmental exposure to electric and magnetic fields at 50 or 60 Hz, and sometimes to electric transients and shocks (normally in the range of a few tens of megahertz).

These dosimeters can record exposure history. This means that every reading can be assigned a specific time allowing separation of the data into categories of occupational exposures. After completing the measurements, the unit shuts itself off and data are kept in a low-power-drain memory. The information can thereafter be retrieved via the port of a desktop personal computer.

Normally, software is provided to display graphs of chronological records of the fields measured at different time resolution levels, total delivered doses and average values, as well as histograms of the data. It is also possible to manipulate histograms to include or exclude specific periods of the day or week, for instance work-hours or work-days.

4

Effects of power frequency electric and magnetic fields

4.1 Electric fields: Distinction between direct and indirect interaction

Exposure of a living organism to electric fields is normally characterized by the unperturbed field strength, that is the field strength measured or calculated with the subject removed from the system. The use of this field is convenient because it involves a quantity that is relatively easy to measure or calculate. The fields that actually act on an exposed organism include E-fields acting on the outer surface of the body and E-fields and current densities induced inside the body. These fields are different from the unperturbed field because of changes caused by the body of the exposed subject. They must, however, be determined in order to specify exposure at the level of living tissues or to relate exposure levels and conditions from one species to another.

Electric fields that act directly on an exposed subject can be categorized as follows:

(a) acting on the outer surface of the body. These fields can cause hairs to vibrate and can thereby be perceived; they may also be able to stimulate other sensory receptors in the skin;

(b) induced inside the body. These fields act at the level of the living cell, and they are accompanied by electric currents through the resistivity of living tissues.

Besides the direct effect of electric fields, secondary short-term effects must be considered when evaluating health risks resulting from electric field exposure. The reason for this is that hazards for some indirect effects are higher than for biological effects due to the direct influence of these fields. The main risks are:

− contact currents which enter a person through electrical conductors coming into contact with the skin;

− when the electrical breakdown potential of air is exceeded, spark discharges introduce transient currents into the body through an arc gap;

− electric or magnetic fields can interfere with the performance of medical implants, e.g. unipolar cardiac pacemakers.

25

In this chapter a distinction is made between effects caused by direct influence of electric fields and indirect effects caused by approaching or touching charged objects, or by electromagnetic interference with electronic implants.

The biological effects (*in vitro* studies, animal experiments) of ELF electromagnetic fields are treated in Appendix A.

4.2 Mechanisms of direct effects of electric fields on the human body

Since the rate of change of ELF electric fields with time is relatively slow, it is useful to first consider an animal or human exposed to a static electric field, E_0. In this case, the effect of E_0 is to induce an electric charge on the surface of the exposed body. This surface charge produces an electric field, E_1, such that (a) E_1 exactly cancels E_0 inside the body, and (b) the total field $E_0 + E_1$ outside the body is enhanced relative to the undisturbed field E_0 at most points on the exterior surface of the body. Since the electric field inside the body is zero, it is clear that the electrical structure inside the body cannot affect the electric fields outside the body. Thus, the only important properties of the body are its shape and location relative to other bodies and the ground.

Now suppose that the external electric field starts to oscillate at an ELF frequency. As the field oscillates, the resulting surface charges will correspondingly oscillate. This requires that electric charges will be continually redistributed on the surface of the body, thereby producing electric fields and currents inside the body.

4.2.1 Surface effects of electric fields

The electric field lines are perpendicular to the surface of the body. A greater concentration of electric field lines (i.e. higher field strength) exists at a curved surface, such as the human head, than on less curved surfaces of the body. For this reason, it is useful to specify the surface electric field that exists on various parts of the body.

The maximum field strength E_{max} at the head of a standing human is $14 \times E_0$ at a frequency of 50 Hz and $18 \times E_0$ at 60 Hz (E_0 is the unperturbed field strength). Figure 13 shows the distortion of the electric field caused by a standing human.

The effects caused by strong surface electric fields include:

(a) stimulation of peripheral receptors in the skin;

(b) vibration of hairs.

The vibration of hairs in an alternating field is shown in figure 14. The amplitude is proportional to the electric field strength, while the frequency of the vibration is twice the field frequency.

Figure 13. Distortion of the electric field by a standing human (50 Hz)

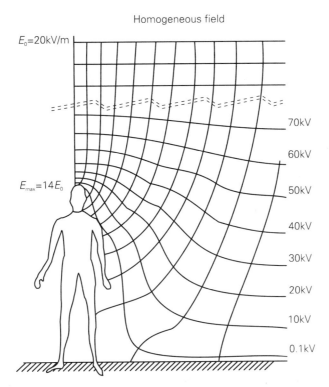

Source: J. Silny: *Wirkung elektrischer Felder auf den Organismus*, Medizin-Technischer Bericht, 1979 (Cologne, Institut zur Erforschung elektrischer Unfälle der Berufsgenossenschaft der Feinmechanik und Elektrotechnik, 1979), p. 8.

4.2.2 Induced electric fields and currents inside the body

Within the body, the two quantities of interest are the total current and the distribution of current density. The total current is more easily measured or calculated, but the current density is more directly relevant in discussion of electric field effects in a particular tissue or organ. Electric field coupling occurs through capacitive and conductive modes.

A body is coupled to an electric field in proportion to its capacitance so that the greater the capacitance the greater the current flow in the body. The capacitance is dependent on the size, especially on the surface area, shape and orientation of the body, so that internal currents will differ between fat and thin persons, between persons standing and reclining, and between persons walking barefoot and those wearing thick rubber-soled shoes or standing on a platform.

A short-circuit current I_{sc} flows in a body placed in an electric field and connected to the ground through a low-resistance path (bare feet, a hand grasping an earthed pole). This current is the sum of all the displacement currents collected over

Figure 14. Vibration of hairs in an alternating electric field

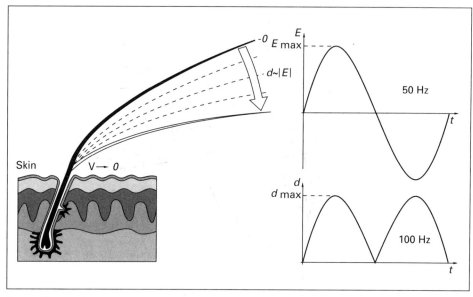

Source: Silny, op. cit., p. 11.

the surface of the body. The only place on the body where the short-circuit current can be measured is where there is connection with the ground. As a rule, a short-circuit current of about 14 µA flows in a body exposed to a vertical electric field of 1 kV/m at a frequency of 50 Hz (16 µA at 60 Hz).

A simple relationship exists between the internal and the external electric field strength, depending on the body part or organ considered and on the exposure conditions. The electrically induced current densities within the body may be calculated by $J = k \cdot f \cdot E$, using suitable k-values (Bernhardt, 1985). A human standing with the longitudinal axis of the body parallel to the external E vector must be considered as "worst case". The values of k can be determined from data from several studies on the absorption at high frequencies within the quasi-static range.

A similar value for k ($3 \cdot 10^{-9}$ S/(Hz m)) was obtained for the cardiac region as well as for the head. For other parts of the body the values of k may be larger, depending also on the exposure conditions, e.g. a threefold value for the neck and a tenfold value for the ankles for a human standing on and in electrical contact with ground (Kaune and Phillips, 1980; Guy et al., 1982; Kaune and Forsythe, 1985, Dimbylow, 1987).

Surface E-field and current-density data derived from Deno's human measurements (Deno, 1977) and Kaune and Phillips (1980) animal data are presented in figure 15. In this figure, the magnitude and frequency of the unperturbed E-field were assumed to be 10 kV/m and 60 Hz, respectively.

Figure 15. Grounded human, pig and rat exposed to a vertical 60 Hz, 10 kV/m electric field

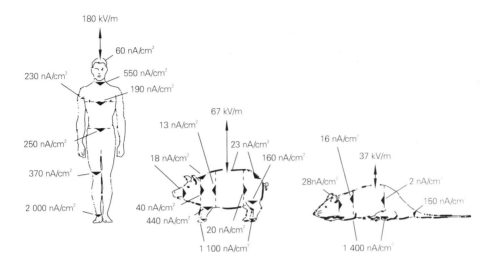

Note: Relative body sizes are not to scale. Surface electric field measurements for human and pig, and surface field estimates for rats, are shown. Estimated axial current densities averaged over selected sections through bodies are shown. Calculated current densities perpendicular to the surface of the body are shown for human and pig.

Source: W. T. Kaune and R. D. Phillips: "Comparison of the coupling of grounded humans, swine and rats to vertical 60 Hz electric fields", in *Bioelectromagnetics*, No. 1, 1980, pp. 117-129, quoted in UNEP/WHO/IRPA: *Environmental Health Criteria 35, Extremely low frequency (ELF) fields* (Geneva, WHO, 1984), fig. 1.

Figure 15 shows that an erect grounded human being (biped) couples more strongly to an ELF electric field than laboratory animals (quadrupeds). At the top of the body surface electric fields are enhanced over the unperturbed field strength present before the subjects entered the field by factors of 18, 6.7 and 3.7 for human beings, pigs and rats, respectively. For an unperturbed field strength of 10 kV/m, average induced axial current densities in the neck, chest, abdomen and lower part of the legs are, respectively: 550, 190, 250 and 2,000 nA/cm^2 for human beings; 40, 13, 20 and 1,100 nA/cm^2 for pigs; and 28, 16, 2 and 1,400 nA/cm^2 for rats.

The difference means that the external unperturbed fields, which are almost always used to specify exposure, must be scaled to equalize internal current densities and E-fields in order to extrapolate biological data from one species to another. For example, a scaling factor for the peak E-field strength acting on the outer surface of the body would be about 4.9 to 1 for humans to rats, while the scaling factor for axial current density in the neck would be about 20 to 1 for the same species comparison. Knowledge must be obtained about the site of action for a particular biological effect before extrapolation of data across species can be accurately and confidently performed.

The analysis of threshold values of electric current densities for different biological effects have shown that:

– for current densities < 1 mA/m^2 no detectable effects have been reported. The average values of the naturally flowing background body current densities are of the same order of magnitude (Bernhardt, 1979);

– current densities of 10 mA/m^2 are the lower end of the range where well-established biological effects are expected and observed (UNEP/WHO/IRPA, 1987). This means that the induced current densities (compare figure 15) are too small to produce significant biological effects. The current densities are very small in comparison with the current densities that flow when contact is made with charged conductors.

Values of the electric field strength for producing different current densities in different body parts are given in table 6.

4.3 Indirect interactions of electric fields

4.3.1 Short-circuit currents

Electric fields at power frequencies can induce an electric charge on ungrounded or poorly grounded metallic objects such as cars, trucks, cranes, wires, fences, and so on. When a person comes in contact with such objects, current to ground flows through the body: the person practically shortcircuits the object.

It is necessary to distinguish between the transient short-circuit current and the steady-state short-circuit current. Zaffanella and Deno (1978) presented data, obtained under various circumstances, that indicated that peak currents of up to several amperes can flow locally for a few microseconds when a person draws a spark discharge from an object.

The steady-state short-circuit current that flows when the charged object is earthed depends on the capacitance of the object to earth, and the open-circuit voltage to which the object is charged, when disconnected from ground, according to the relation: $I_{sc} = \omega V_{oc} \cdot C$, where I_{sc} is the short-circuit current, ω is the angular frequency of the electric field, V_{oc} is the open-circuit voltage, and C the capacitance of the object to earth.

Typical capacitances for objects range from 700 pF for a small vehicle to several thousand pF for buses and large trucks and about 1,000 pF for a 150 m long fence. Thus, the short-circuit current for a 150 m fence could be as great as 2.2 mA, if the fence were located in a field of 5 kV/m. Zaffanella and Deno (1978) measured the short-circuit currents of a farm tractor, a jeep wagon and a school bus. In a 10 kV/m electric field, these vehicles conducted 0.6, 1.1 and 3.9 mA of current to earth, respectively. Although the transient currents are of appreciable magnitude, they should not present a hazard if appropriate safety procedures are followed. Good engineering practices to reduce the risk of shocks include the careful earthing of fences, gutters and other long metallic objects in a strong electric field.

Table 6. Induced current density ranges at 50 or 60 Hz for producing biological effects

Electric field strength (kV/m)				Current density (mA/m²)	Effects	Magnetic flux density (mT)		
Trunk (average)	Head	Neck	Ankles, when both feet are grounded			Trunk ($R = 0.3$ m)	Head ($R = 0.075$ m)	Wrist/ankles ($R = 0.03$ m)
> 300	> 1000	> 200	> 50	> 1000	Cardiac extrasystoles and ventricular fibrillation possible; definite health hazards	> 60	> 250	> 600
				100–1000	Changes in central nervous system excitability established; range, where stimulation of excitable tissue is observed; possible health hazards	6–60	25–250	60–600
30–300	100–1000	20–200	5–50	10–100	Well-established effects, evident visual (magneto-phosphenes) and possible nervous system effects; induction of bone reunion reported	0.6–6	2.5–25	6–60
3–30	10–100	2–20	0.5–5	1–10	Minor biological effects reported	0.06–0.6	0.25–2.5	0.6–6
< 3	< 10	< 2	< 0.5	< 1	Absence of well-established effects	< 0.06	< 0.25	< 0.6

Left: Values of the electric field strength for approximately producing these current densities in different body parts.
Right: Values of the magnetic flux density for approximately inducing these current densities in peripheral regions of different body parts (a homogeneous conductivity of 0.25 S/m is assumed).
Source: UNEP/WHO/IRPA, 1987.

There are several reactions and sensations that are of interest in assessing the coupling effects of power frequency electric fields (IEEE, 1978). The following list is in order of increasing stimulus:

Perception: The person is just able to detect the stimulus at a given detection probability. Because the current density within the body is the important parameter, there is a difference in the current perception threshold when comparing touch with grip perception.

Annoyance: The person would consider the sensation to be a mild irritant if it were to occur repeatedly.

Startle: If the stimulus occurred unexpectedly, it would most likely produce an unintentional muscular reflex capable of being hazardous under a defined set of circumstances.

Aversion: If a person receives one exposure, he or she would be motivated to avoid situations that would lead to a similar experience.

The sensations that result from microshocks are not hazardous, but the annoyance they cause may be highly significant in the evaluation of effects attributed to the fields. The following reactions are associated only with currents from direct contact:

Let-go: A person cannot let go of a gripped conductor as long as the stimulus persists due to uncontrollable muscle contraction. If a person is exposed to prolonged currents somewhat above the let-go level that pass through the chest, breathing becomes difficult and, eventually, the person may become exhausted and die. For this reason, let-go currents would be considered as potentially lethal in prolonged exposure.

Respiratory tetanus: A person is unable to breathe as long as the stimulus is applied due to contraction of the muscles responsible for breathing.

Fibrillation: Uncoordinated asynchronous heart contractions which produce no pumping action.

Table 7 summarizes experimental and calculated threshold levels for effects of currents passing through the human body. The values are for 50 per cent of persons and are broadly classed as perception, startle, painful shocks, let-go and respiratory tetanus. With regard to perception thresholds, it should be mentioned that currents considerably below the levels of table 7 can be perceived if the contact is with the tongue or an open wound.

Short-circuit currents from different objects to ground were calculated by Guy and confirmed by measurements (Guy and Chou, 1982, Guy, 1985). The measurements indicated that a current equal to the short-circuit current flows through the

EFFECTS OF ELECTRIC AND MAGNETIC FIELDS

body of a person touching the object when the person is barefoot on wet ground (worst case). The current is reduced when the person wears shoes. By using the data of Guy and the threshold levels of table 7, values for electric field strengths producing an effect when a grounded person touches an isolated object in an electric field are summarized in table 8. For comparison, the lower part of table 8 gives perception thresholds of short-circuit body currents of grounded men and women in an electric field.

Table 7. Effects of currents passing through the human body[a]

Effects	Subject	Current in mA	
		50 and 60 Hz	300 Hz
Touch perception	Men	0.36	(0.47)
(finger contact)	Women	0.24	(0.31)
	Children	*0.18*	*0.24*
Grip perception	Men	1.1	1.3
	Women	0.7	*0.9*
	Children	*0.55*	*0.65*
Shock, not painful	Men	1.8	(2.3)
(grasping contact)	Women	1.2	*1.5*
	Children	*0.9*	*1.1*
Pain,	Men	(1.8)	(2.4)
(finger contact)	Women	*1.2*	*1.6*
	Children	*0.9*	*1.2*
Shock, painful, muscle control	Men	9	(11.7)
(let-go threshold for 0.5% of population)	Women	6	*7.8*
	Children	4.5	*5.9*
Painful shock, let-go threshold	Men	16	18
	Women	10.5	*12*
	Children	*8*	*9*
Severe shock, breathing difficult	Men	23	(30)
	Women	15	*20*
	Children	*12*	*15*

[a] Experimental data for 50 per cent of men, women and children (from Dalziel, 1954, 1968; Deno, 1974; Guy and Chou, 1982, Guy, 1985; Chatterjee et al., 1986). Data in brackets were calculated by using the frequency factors for perception thresholds, for pain and let-go thresholds, given in IEC Publication 479. Data in italics were calculated by assuming thresholds for women two-thirds of those of men and thresholds for children one-half of those of men.

Sources: IEEE, 1978; Guy, 1985.

33

PROTECTION FROM POWER FREQUENCY ELECTRIC AND MAGNETIC FIELDS

Table 8. Electric field strength for person-to-vehicle currents and short-circuit body current for persons exposed to electric fields with feet grounded

Effect	Subject	Field strength (kV/m)			
		$16^2/_3$ Hz	50 Hz	60 Hz	300 Hz
Indirect effects:					
Touch perception	Men	14.6	5.0	4.0	1.1
(finger contact, car)	Children	7.3	2.5	2.0	0.54
Pain, finger contact (car)	Men	73	24	20.4	5.4
	Women	49	16	13.6	3.6
	Children	36	12	10.2	2.7
Painful shock, car contact	Children	183	61	51	13
(let-go threshold, 0.5 %)					
Painful shock, truck contact	Children	28	9.4	7.8	2.0
(let-go threshold, 0.5 %)					
Direct effect:					
Perception of short-circuit	Men	253	85	71	17
body current of grounded	Women	182	61	51	13
subjects					

Sources: table 7; Deno, 1974; Guy, 1985.

4.3.2 Reactions to spark discharges

Capacitive spark discharges may occur between a person in an electric field and a conducting object. The person acts as capacitor on which a charge is induced by the action of the electric field. When the person and the conducting object are a fraction of a millimeter apart and the voltage potential difference is larger than about 500 V, then the electric breakdown potential of air may be exceeded and an electric arc results.

These discharges are qualitatively similar to the static discharges that can occur when a person walks across a carpet on a dry day, except that a carpet spark is a single transient event whereas an alternating current (AC) spark discharge is continued as long as a small air gap is maintained. If the gap is closed the spark discharge is replaced by a contact current.

Frequently the current passing in an easily felt spark discharge may be below the perception level for direct contact currents, such as when gripping a conductor. The distinction is due primarily to the way current is distributed on the skin. The current for direct contact is generally more dispersed in skin area than with spark discharges, where the current may be introduced through a very small area resulting in a large value for the current density.

The person's reaction to an AC spark discharge (time-constant several μs, Reilly and Larkin, 1983) is dependent on the open-circuit voltage developed on the charge-

34

EFFECTS OF ELECTRIC AND MAGNETIC FIELDS

collecting object as well as on its capacitance to ground. The perception energy thresholds for spark discharges (male, fingertip or thumbtip contacts) are 0.07 mJ for a humid environment and 0.35 mJ for a dry environment (IEEE, 1978). The threshold energy for unpleasant painful shocks is about 0.5 to 1.5 mJ (Guy, 1985). Threshold data for a person acting as a charge-collecting object are summarized in table 9 (Zaffanella and Deno, 1978).

Table 9. Direct and indirect effects of 50 and 60 Hz electric fields

Field strength (kV/m)	Effect
>50	Direct perception for short-circuit body current of grounded man
20-24	Median pain perception for men, finger contact, car
20	Perception threshold for 50% of men with sensations on their head, head hair or tingling between body and clothes
16-20	0.5% let-go threshold for men, truck contact
14-16	Median pain perception for women, finger contact, car
11.5-14	0.5% let-go threshold for children, bus contact
11-13	0.5% let-go threshold for women, truck contact
10-12	Median pain perception for children, finger contact, car
8-10	0.5% let-go threshold for children, truck contact
4-7	Median annoyance level for spark discharges, the person acting as charge-collecting object (170 pF)
4-5	Median touch perception for men, finger contact, car
2.5-6	90% perception levels for spark discharges, the person acting as charge-collecting object (170 pF)
3	Perception threshold for 5% of men with sensations on their head, head hair, or tingling between body and clothes
2-2.5	Median touch perception for children, finger contact, car
2.5	Threshold for interference with extremely sensitive unipolar cardiac pacemakers (0.5 mV peak to peak voltage sensitivity)
1.2-2.5	Median perception levels for spark discharges, the person acting as charge-collecting object (170 pF)
0.6-1.5	10% perception levels for spark discharges, the person acting as charge-collecting object (170 pF)

Sources: IEEE, 1978; Zaffanella and Deno, 1978; UNEP/WHO/IRPA, 1984.

4.4 Mechanisms of interaction of magnetic fields

4.4.1 Basic mechanisms of interaction

The most important physical mechanisms by which magnetic fields interact with living matter may be summarized as follows (UNEP/WHO/IRPA, 1987; Tenforde, 1985): At the atomic and subatomic levels, several types of magnetic field interactions have been shown to occur in biological systems. Two such interactions are the nuclear magnetic resonance in living tissues and the effects on electronic spin states and their relevance to certain classes of electron-transfer reactions. At the level of macromolecules and larger structures, interactions of magnetic fields with biological systems can be characterized as electrodynamic or magnetomechanical in nature. Electrodynamic effects originate from the interaction of magnetic fields with moving conductors such as electrolyte flows, leading to the induction of electrical potentials and currents. Magnetic fields which are time varying also interact with living tissues at the macroscopic and microscopic levels to produce circulating currents via the mechanisms of magnetic induction.

From the analysis of established mechanisms of magnetic field interaction with living matter it can be concluded that for ELF magnetic fields the induction is predominant. The appropriate dosimetric quantity is the induced electric field strength at the cellular level in the living tissue or – connected with the specific conductivity of the medium – the induced current density (Bernhardt, 1979, 1985). By comparing the current densities, it may be possible to predict effects in human beings from those found in studies on animals and isolated cells. In this context, it is irrelevant whether the current density surrounding a cell is introduced into the body through electrodes or induced in the body by external magnetic fields. However, the current paths within the body may be different. The evaluation of human exposure using current densities is based primarily on a concept of "dose" to the critical organs. In addition the parameters of internal field strength and duration should be taken into account. Basic protection limits can be expressed in current densities; derived protection limits can be expressed as exposures to external magnetic fields, where field strength, frequency, orientation of the body and duration of exposure need to be specified. Refinements may include field gradient values, partial body exposure, and so on.

4.4.2 Induced electric fields and current densities

There is very little experimental or theoretical work dealing with the coupling of magnetic fields of ELF frequencies to models of living organisms. Spiegel (1977) described magnetic field coupling on spherical models, and Gandhi et al. (1984) calculated induced current densities in the torso of a human using a multidimensional lattice of impedance elements. However, using "worst case" assumptions, an estimate of the order of magnitude for "safe" and "dangerous" magnetic field strengths and their frequency dependence is possible. Considering the cardiac region and the brain as "critical" organs, rough "worst-case" calculations were made (Bernhardt, 1979, 1985). Differences in conductivity of the white and the grey cerebral substance, and the anisotropic nature of conductivity at frequencies below approximately 10 kHz, were left out of consideration. For sinusoidal fields, the current densities (peak values) in peripheral regions can be calculated from $J = \pi \cdot R \cdot \sigma \cdot f \cdot B_0$ (R: radius of the current loop; σ: conductivity; f: frequency; B_0: peak value of magnetic flux density). Because of the uncertainties of the current loops and of the values for the conductivities, these calculations are only rough estimates. Values of the magnetic flux density for inducing current densities in peripheral regions in different body parts are given in table 6.

A potentially important target of ELF magnetic field interactions is the nervous system. From a consideration of the naturally occurring fields in the central

EFFECTS OF ELECTRIC AND MAGNETIC FIELDS

nervous system, it can be concluded that magnetic fields in the 1-100 Hz frequency range, capable of inducing current densities in tissue of approximately 1 mA/m^2 or smaller, should not have a direct effect on the brain's electrical activity (Bernhardt, 1979). Induced fields sufficient to exceed a threshold depolarization value can result in an action potential. These effects are well understood. ELF magnetic fields inducing such large depolarizations may result in nerve stimulation or muscle contraction, or even in fibrillation.

With regard to "hazardous values" and the definition of field strengths that lead to injury, the most serious injury may be the initiation of heart fibrillation. The threshold for extrasystole induction at 60 Hz is estimated to be above 300 mT for stimulation times of 1 second or longer, and the threshold for ventricular fibrillation is 1 to 1.5 T (Bernhardt, 1985). For shorter exposure times, higher field strengths are necessary to produce similar biological effects.

4.5 Electromagnetic interference (EMI) with cardiac pacemakers

An implanted pacemaker is an electromedical device that artificially stimulates the heart, thus making it possible for persons with certain heart diseases to lead relatively normal lives. Certain types of modern cardiac pacemaker exhibit malfunction in response to EMI produced either by endogenous muscle potentials or by external sources such as high-current systems. Two different configurations of electrode leads are used in pacemakers, and these have very different sensitivities to EMI. In the first type, termed the "bipolar" design, both leads are implanted within the heart at a typical separation distance of 3 cm. In the second type, termed the "unipolar" design, the cathode lead is implanted in the heart and the pacemaker case serves as the anode.

For *electric field* interference, unipolar pacemakers are especially sensitive, because they have only one electrically isolated electrode and the field can induce an electric voltage between electrode tip and pacemaker.

For 50 Hz electric fields, direct functional impairments of presently implanted pacemakers are not expected for field strengths below 2.5 kV/m (UNEP/WHO/IRPA, 1984).

For *magnetic field* interference, unipolar pacemakers are again especially sensitive, because they have only one electrically isolated electrode, which can open an area up to about 600 cm^2 through which a magnetic field can induce an electric voltage between electrode tip and pacemaker. The probability that a malfunction will occur in the presence of an external magnetic field is strongly dependent on the pacemaker model, on the value of the programmed sensing voltage and on the area of the pacemaker loop which is laid during implantation.

Assuming pacemaker sensitivities of 0.5 to 2 mV for 50 Hz, interference field strengths (peak values) of 11 to 44 A/m (13.8 to 55 µT) may be calculated. Similar results were obtained by Bridges and Frazier (1979). Measurements with 20 different pacemakers have shown (Matthes and Bernhardt, 1988) that the values for the magnetic field strength for interference were between 10 and 100 A/m for the worst case at 50 Hz. For a more realistic effective area of the pacemaker electrode of 150 cm^2, it can be concluded that some sensitive pacemakers may be influenced at field strengths of about 40 A/m (50 μT), and almost all pacemakers are influenced at field strengths of 200 A/m (250 μT).

Pacemaker malfunctions produced by power-frequency magnetic fields require field levels that are greater than those associated with high-voltage transmission lines and most other types of electrical system. However, the fields in the immediate vicinity of various types of industrial machinery and appliance are sufficiently strong to represent a potential source of EMI that could alter pacemaker functioning.

5

Human health effects of power frequency fields

5.1 Introduction

The effects of electric and magnetic fields at the power frequencies of 50-60 Hz on human health can be divided into effects of an acute nature that occur during the exposure to such fields and disappear after the cessation of exposure, and effects which could occur after prolonged exposures or after a long delay.

In the category of acute effects, it is clear that such effects tend to be largely reversible and they can be studied fairly directly because of the temporal association between exposure and effect. This close temporal association also makes it possible to establish the relationship between the strength of the electric or magnetic fields and the magnitude of the observed and reported effects. An additional advantage is that in such studies many if not most of the exposed persons will report the same effects. Acute effects of ELF fields were reviewed recently by Gamberale (1990).

The second category of effects that have been considered is that of delayed and stochastic effects which include excess risk of leukaemia and brain cancer, and excess reproductive failures. Epidemiological studies of the possible association of exposure to power frequency fields and the delayed outcomes are difficult to design, execute and interpret. The difficulties arise because exposures are not well described and are hard to estimate over the number of years for which they need to be known. In addition, associations of this type are not reported with much consistency, and when they are reported they tend to be weak associations. It is usually difficult to assess the risks contributed by simultaneous exposures to other known risk factors. The current state of knowledge does not permit an estimate of the risk which could be associated with ELF exposures, even if one should exist. As a result there are no known guidelines or standards that address the second category of effects.

5.2 Studies of acute effects of ELF fields

5.2.1 Observational studies of acute effects

As has been discussed in the preceding chapters, some occupational environments are characterized by intense electric and magnetic fields that are likely to cause localized electric fields at the skin and inside the human body and cause induced currents to arise throughout the body. When such fields are strong enough, the effects in the body are such that the senses can perceive them and there will be an awareness of the presence of the field. It is also possible to assess the magnitude of the electric and magnetic fields that give rise to the perceptions and other acute reactions. In observations at the actual workplace such exposures are usually not controlled by the investigator, but they can be described. In studies with volunteers in laboratory settings the exposure can be controlled and chosen very precisely, and possible simultaneous exposures to other factors such as solvents and other chemicals can be precluded.

Human subjects exposed to a vertical E-field were able to detect hand hair vibration at levels 1 m above ground of 1.5-2 kV/m, when standing and holding hands over their heads. Median values of perception by hand hair vibration are about 7 kV/m (1 m above ground) and about 20 kV/m for head hair. A tingling sensation was reported at fields above 5 kV/m, and annoyance was registered from fields of about 10 kV/m at 1 m above ground level (Deno and Zaffanella, 1982; UNEP/WHO/IRPA, 1984, 1987; Cabanes and Gary, 1981; Reilly et al., 1981).

If the ELF magnetic field exceeds 10 mT at frequencies above 10 Hz, there are manifestations of flickering light perception which are referred to as magnetophosphenes (Lövsund, 1981). Some animals can perceive the earth's magnetic field and use it for orientation. Such orientation can be disturbed by ELF magnetic fields of sufficient strength, but similar reactions have not been described for humans or other animals. Electronic pacemakers, depending on their design and construction, have some vulnerability for malfunction caused by currents that can be induced in their electronics or the leads by external electric and magnetic fields (Bridges and Frazier, 1979).

A number of studies have reported that exposure to electric and magnetic fields in switch-yards was associated with subjective and objective adverse effects (Asanova and Rakov, 1966; Sazanova, 1967; Revnova et al., 1968; Fole and Dutrus, 1974). Subjective results were complaints by workers of excess headaches, dizziness, nausea, fatigue, irritability and other symptoms. There were also reports of objective changes in haematology, and in electrocardiograms (ECG) and electroencephalograms (EEG).

A large number of other investigators have made observations in other settings and have been unable to confirm the earlier reports (Singewald et al., 1973; Knave et al., 1979; Broadbent et al., 1985; Baroncelli et al., 1986). In addition, when non-occupational exposures of the general public living in close proximity to transmission lines were evaluated, such studies were not able to find adverse health effects in these populations.

In a recent study (Gamberale et al., 1989), workers were observed very closely, with medical examinations before and after a work-day of simulated inspections of insulators on a transmission line. There was one day of such inspections with the line energized at 400 kV, and another day with the same work without the line being energized. Blood samples were collected from 26 workers during these two work-days and there was no statistically significant difference in any of a large number of blood levels evaluated, indicating that there were no differences in the acute effects of performing these tasks with the line energized or de-energized. The mean exposure

HUMAN HEALTH EFFECTS

for the energized day was 2.8 kV/m and 23.3 microtesla or 233 milligauss. It is clear that such studies cannot address the question of delayed effects of repeated workplace exposures at these levels.

5.2.2 Studies of volunteers under controlled conditions

Studies of volunteers under laboratory conditions allow observations to be made of the effects of precisely defined exposures and of the responses produced in immediate association with such exposures.

In controlled exposure of volunteers, Filipov observed a threshold for haematological changes, and he suggests that none would occur in fields below 5 kV/m (Filipov, 1972).

A number of studies (Hauf, 1982; Sander et al., 1982; Fotopoulos et al., 1987; Graham et al., 1987) have exposed volunteer subjects under controlled conditions to electric fields in the 1-36 kV/m range, and to magnetic fields in the range of 19 microtesla to 5 millitesla without significant changes in a number of blood levels, ECG and EEG, and levels of performance. In some of the studies there were field-related changes in evoked potentials following acoustic or visual stimulation, as well as a reduction in the resting heart rate (Hauf, 1982; Sander et al., 1982). At very high fields of 36 kV/m there have been reports of changes in mood and reduction of reasoning performance (Stollery, 1986).

It is difficult to arrive at very firm conclusions based on these reports of studies on acute health effects of electric and magnetic fields. It is probably safe to say that if any such effects actually occur then they are not severe or they may be associated with other factors simultaneously present in the environment. These studies are also difficult to evaluate in that the mix of electric field exposure and magnetic field exposure is not always given.

5.3 Occupational epidemiological studies of delayed effects

A number of the studies of acute effects referred to in the previous section used epidemiological methods in the form of cross-sectional studies of populations working in switch-yards or on high-voltage transmission lines. Such studies are not able to detect effectively the presence or suggest the absence of health effects which require a long period of time to develop. Any association between ELF electric or magnetic fields and delayed effects such as brain cancer or leukaemia will express itself in a small increase in the occurrence of relatively rare events. Epidemiological methods that address such problems include cohort studies in which a large number of people are followed up for a long period of time, or case-control studies in which cases that developed in a defined population are compared with controls from the same defined population. These studies ideally should have complete information about the exposure in both groups, and should be large enough to have adequate statistical power.

Studies which have been conducted to investigate an association between ELF field exposure and excess cancer often have difficulties describing the exact exposure over many years for many individuals. With very few exceptions occupational studies, which have only been conducted in the last dozen years, have relied on job categories as an indicator for exposure. As a result, it is not possible to distinguish between the effects of exposure to ELF fields and all other factors in that occupational environment which might have contributed to the outcome in question.

The first of a series of studies linking occupation involving electric and magnetic field exposure to excess cancer was by Milham (1982). Earlier, Wertheimer and Leeper (1979) reported an increased risk of childhood cancer in children in households with an electric supply wiring configuration that would expose them to an elevated level of 60 Hz magnetic fields.

Since these earlier studies a large number of additional reports of occupational exposures to ELF fields and the possible incidence of different forms of cancer have been carried out. These studies have been conducted under different conditions, mostly without quantitative characterization of the exposure. They have also considered various outcomes such as different forms of leukaemia, different forms of brain cancer, and so on. Many of these studies have been reviewed by Calle and Savitz (1985), Coleman and Beral (1988), Silverman (1990) and Dennis et al. (1991). Such reviews point out that there are a number of studies suggesting that those employed in electrical occupations are at a small excess risk of leukaemia, more particularly acute myeloid leukaemia.

It is difficult to draw firm conclusions based on the epidemiological studies since the findings cannot be considered conclusive and do not persuade most investigators that a causal relationship has been proven. On the other hand, the findings do not allow a conclusion that such associations are not causally related. Further work is required. It is important to note that in almost all the occupational studies, the exposure to electromagnetic fields was assumed and not measured. At the same time, exposures to other risk factors in the same occupational environment are not usually evaluated, so that confounding effects of such factors cannot be excluded.

It is not possible to use the inconclusive results of the epidemiological studies in any health risk assessment (NRPB, 1992). Thus limits of electric and magnetic field exposure cannot be based on these studies.

6

Occupational exposure limits

6.1 The basis for setting protection standards

In recent years, the ever-growing use of electrical appliances and the consequent demand for electric power have greatly increased the awareness, both in the scientific community and among the general public, of possible health risks of electric and magnetic fields at 50/60 Hz frequencies. Particular concern has been raised about power lines. The request of workers for assurance about health risks has increased to the point that some regulatory agencies have been confronted with the problem of issuing standards on the exposure levels to 50/60 Hz electric and magnetic fields.

According to a generally accepted definition (UNEP/WHO/IRPA, 1984; Repacholi, 1985), a standard is a general term indicating a set of specifications or rules to promote the safety of an individual or group of people. Standards can be divided into regulations and guidelines. A regulation is a mandatory standard promulgated under a legal statute, whereas a guideline is in general a set of recommendations issued for guidance only. A standard can specify any kind of rule to be complied with in order to reduce health risks within acceptable limits. In the case of electromagnetic fields, such specifications include:

– field limits, i.e. maximum levels that the electric or magnetic field may attain in specific points or regions;

– exposure limits, i.e. maximum levels to which exposure of the whole body or part of it is allowed;

– prohibition or limitation of access to specific areas;

– rules for the siting of an industrial facility;

– details on the performance, construction, design or functioning of a device.

A limit should be based on the ascertainment of a threshold value for adverse health effects; in the absence of any apparent threshold, a level of acceptable risk should be determined, based on a risk-benefit analysis, which in turn requires some correspondence to be established between different exposure levels and biological or health effects. The definition of such limits is the basis for promulgating protection standards.

As pointed out by several authors (Grandolfo et al., 1985; Tenforde and Kaune, 1987), data on biological effects of power frequency fields and their mechanisms of interaction with cellular and tissue systems are at present controversial. Partly for this reason only a few standards have so far been issued.

At present, the greatest concern is for high-voltage transmission lines, and sometimes it gives rise to protests and strong opposition towards the installation of new facilities. The intervention of objectors at different levels of the decision-making bodies may even cause long delays in the licensing procedure.

Taking into account the economic and social cost of the above-mentioned delays, and sometimes of legal actions, it may be useful also for the electric utilities to define limit values, provided that they are technically reasonable and scientifically valid. This opinion has been expressed, for example, by the Committee for Medical Studies of the International Union of Producers and Distributors of Electrical Energy (UNIPEDE, 1985).

6.2 International recommendations

Following the recommendations of the United Nations Conference on the Human Environment held in Stockholm in 1972, and in response to a number of World Health resolutions and the recommendation of the Governing Council of the United Nations Environment Programme (UNEP), a programme on the integrated assessment of the health effects of environmental pollution was initiated in 1973. The programme, known as the WHO Environmental Health Criteria Programme, has been implemented with the support of the UNEP Environment Fund. In 1980 the Environmental Health Criteria Programme was incorporated into the UNEP/ILO/WHO International Programme on Chemical Safety (IPCS).

The International Radiation Protection Association (IRPA) initiated activities concerned with non-ionizing radiation by forming a Working Group on Non-Ionizing Radiation in 1974. This Working Group later became the International Non-Ionizing Radiation Committee (INIRC) at the IRPA meeting in Paris in 1977. In 1992, the IRPA/INIRC became an independent scientific body called the International Commission on Non-Ionizing Radiation Protection (ICNIRP) and has the same responsibility for NIR as the ICRP has for ionizing radiation. The IRPA/INIRC (now the ICNIRP) reviews the scientific literature on non-ionizing radiation, makes assessments of the health risks of human exposure to such radiation and, in cooperation with the Environmental Health Division of the WHO has undertaken responsibility for the development of environmental health criteria documents on non-ionizing radiation.

For each kind of radiation the criteria document includes an overview of the physical characteristics, measurement techniques and instrumentation, sources and applications, and finally of the scientific literature on biological and health effects. These criteria documents become the scientific bases for the development of guidelines on exposure limits and guidance for safe practice.

Two joint WHO/IRPA Task Groups on Environmental Health Criteria Documents for Extremely Low Frequency Fields (ELF) (UNEP/WHO/IRPA, 1984) and Magnetic Fields (UNEP/WHO/IRPA, 1987) reviewed and analysed the existing scientific literature. They made an evaluation of the health risks of exposure to electromagnetic fields, considered rationales for the development of human exposure limits, and gave advice and recommendations for further research, so that on the basis of more definite data a critical revision of the existing standards can take place at an international level. The WHO/IRPA Task Group on ELF (UNEP/WHO/IRPA, 1984) concluded that:

- whilst it would be prudent in the present state of scientific knowledge not to make unqualified statements about the safety of intermittent exposure to electric fields, there is no need to limit access to regions where the field strength is below about 10 kV/m. Even at this field strength, some individuals may experience uncomfortable secondary physical phenomena such as spark discharge, shocks or stimulation of the tactile sense;

- furthermore, it is not possible from present knowledge to make a definitive statement about the safety or hazard associated with long-term exposure to sinusoidal electric fields in the range of 1-10 kV/m;

- in the absence of specific evidence of particular risk or disease syndromes associated with such exposure, and in view of experimental findings on the biological effects of exposure, it is recommended that efforts be made to limit exposure, particularly for members of the general population, to levels as low as can be reasonably achieved.

IRPA/INIRC (1990) issued interim guidelines which apply to human exposure to electric and magnetic fields at frequencies of 50 or 60 Hz.

The basic criterion is to limit current densities induced in the head and trunk by continuous exposure to power frequency fields to no more than about 10 mA/m^2.

A summary of the limits recommended by IRPA/INIRC for occupational exposures is given in table 10.

Continuous occupational exposure during the working day should be limited to rms unperturbed electric field strengths not greater than 10 kV/m.

Short-term occupational exposure to rms electric field strengths between 10 and 30 kV/m is permitted, provided that the rms electric field strength (kV/m) times the duration of exposure (hours) does not exceed 80 for the whole working day.

Table 10.   Limits of occupational exposure to 50/60 Hz electric and magnetic fields

Exposure characteristics	Electric field strength (kV/m (rms))	Magnetic flux density (mT (rms))
Whole working day	10	0.5
Short term	30[a]	5[b]
For limbs	–	25

[a] The duration of exposure to fields between 10 and 30 kV/m may be calculated from the formula $t \leqslant 80/E$, where t is the duration in hours per work-day and E is the electric field strength in kV/m.
[b] Maximum exposure duration is two hours per work-day.
Source: IRPA/INIRC, 1990.

Continuous occupational exposure during the working day should be limited to rms magnetic flux densities not greater than 0.5 mT.

Short-term occupational whole-body exposure for up to two hours per workday should not exceed a magnetic flux density of 5 mT. When restricted to the limbs, exposures up to 25 mT can be permitted.

The IRPA/INIRC exposure limits are based on established or predicted effects of exposure to 50/60 Hz fields. Although some epidemiological studies suggest an association between exposure to 50/60 Hz fields and cancer, others do not. Not only is this association not proven, but present data do not provide any basis for health risk assessment useful for the development of exposure limits.

Current laboratory studies are testing the hypothesis that 50/60 Hz fields may act as, or with, a cancer promoter. These studies are still exploratory in nature and have not established any human health risk from exposure to these fields.

These limits have been developed from present knowledge, but there are still areas of research where questions have been raised that need to be addressed. A major research effort is required to supplement our knowledge on the health consequences, if any, of long-term continuous exposure of humans to low level 50/60 Hz fields.

There is an ever-increasing number of people wearing implanted cardiac pacemakers which may be sensitive to interference from electric and magnetic fields. These people may not always be adequately protected against interference at some of the exposure limits recommended by IRPA/INIRC.

7

Control of, and protection from, exposure to 50 and 60 Hz electric and magnetic fields

7.1 General considerations and scope

As indicated in Chapter 2, three kinds of situations are identified, namely:

– common use of electricity,

– use of specific electrical equipment,

– high-voltage transmission lines.

In the first case, good practice and electrical safety should be applied for the prevention of electrical accidents. Information and training should be provided. However, no preventive measure is necessary or useful as regards the protection against electric and magnetic fields associated with the common use of electricity.

The use of specific electrical equipment and practices such as induction heating, arc furnaces and welding machines needs to be subject to control since they may give rise to significant exposure to electric and magnetic fields. Such equipment and processes should be identified by the competent authority.

Regulations and standards should be adopted to control exposures from such sources and practices including the establishment, as appropriate, of a system of notification, registration and licensing.

In the case of high-voltage transmission lines, a system of control and protection including procedures and safe practice requirements is already established. It should include the provisions necessary to ensure a suitable degree of protection against power frequency electric and magnetic fields.

The provisions of this chapter provide practical guidance on the programmes of prevention which should be introduced to control exposures and to ensure protection in cases where there are occupational hazards due to power frequency electric and magnetic fields.

The responsibilities for the protection of workers and the general public against the potentially adverse effects of exposure to 50/60 Hz electric and magnetic fields should be clearly assigned.

General rules on health surveillance have been established by the International Labour Office in the Occupational Health Services Convention (No. 161), and Recommendation (No. 171), of 1985.

7.2 Role of the competent authorities

The competent or regulatory authorities whose terms of reference and functions are concerned with protection against the harmful effects of power frequency electric and magnetic fields should cooperate with each another. This cooperation is necessary to ensure that each authority, or department within an authority, is well aware of the responsibilities of the other organs so as to avoid duplication of effort.

The competent authorities should formulate the necessary regulations (based on the national standards described in Chapter 6) for protection against power frequency electric and magnetic fields. This should be undertaken in consultation with the representative organizations of the employers and workers concerned. In addition, the authorities should provide, as necessary, detailed specific guidance for the design, manufacture and use of the various sources of electric and magnetic fields. Such regulations should also specify the relevant rules, standards and occupational exposure conditions.

The classification of sources of exposure or practices according to notification, registration or licensing requirements should be based on the health hazards involved. Whenever more rigid control is necessary, then the authorities should specify the exposure sources requiring a licence, together with the procedures and conditions for obtaining one.

The licence permit should comprise follow-up methods to ensure compliance during plant construction, commissioning and operation, including future modification of design or working procedures. It should be remembered that the granting of a licence should not preclude a change during the period of its validity. A request for such a change may be initiated by the licensee, or the change may be desirable or necessary as a result of experience gained during operation at the workplace or elsewhere, or as a consequence of technological innovation or of safety research and development.

The licensing process should define the various responsibilities concerned with the planning, design, construction, commissioning and operation of the facility. The licensee should submit and make available to the competent authorities all the information requested. Whenever a significant change in operation or use of the exposure source occurs that is relevant to the licensing process, the licensee should submit that information to the authorities, preferably according to a specified procedure.

The competent authorities should establish a system of inspection to supervise safety precautions in order to ensure compliance with the relevant standards and with the requirements as specified in the licence. They should also assume the necessary power to intervene in cases of non-compliance with the standards. Any situation which has resulted or is expected to result in field levels in excess of the exposure limits should be reported in an approved manner.

In summary, it is recommended that the competent authorities consider the following steps:

(a) development and adoption of exposure limits and the implementation of a compliance programme;

CONTROL OF EXPOSURE TO 50 AND 60 Hz FIELDS

(b) development of technical standards to reduce the susceptibility to electromagnetic interference, e.g. for pacemakers;

(c) development of standards defining zones with prohibited or limited access around sources of strong electric and magnetic fields because of electromagnetic interference (e.g. for pacemakers and other implanted devices). The use of appropriate warning signs should be considered;

(d) requirement of specific assignment of responsibility for the safety of workers and the public to a person at each site with high exposure potential;

(e) drafting of guidelines or codes of practice for worker safety in 50/60 Hz electromagnetic fields;

(f) development of standardized measurement procedures and survey techniques;

(g) requirements for informing workers on the effects of exposure to 50/60 Hz fields and the measures and rules which are designed to protect them.

7.3 Responsibility of the employer

The owner of the device or installation producing power frequency electric and magnetic fields:

(a) is responsible for the safety of employees;

(b) is responsible for purchasing or providing the equipment which meets all appropriate governmental standards when new and during the time of use;

(c) is responsible for assuring that the ELF equipment meets appropriate safety standards and safety requirements as specified in this book;

(d) is responsible for action to reduce the exposure of workers to ELF fields and for the organizational arrangements required to prevent the risks associated with exposure;

(e) should establish and publicize (in writing) a general policy emphasizing the importance of prevention, and should take the decisions and the practical steps required to give effect to national regulations and to this book.

Responsibility may be delegated by the owner depending on the size of the organization and the amount and strength of ELF fields from the equipment used. Without prejudice to the responsibility of each employer for the health and safety of the workers, and with due regard to the necessity for the workers to participate in matters of occupational health and safety, one or more persons may be designated to carry out the role of responsible user and safety inspector.

7.4 Duties of workers

Workers in charge of the day-to-day operation and maintenance of devices producing power frequency electric and magnetic fields must:

(a) understand the hazards associated with operating the specific devices assigned to them and, in particular, the importance of:

   (i) any interlock systems and dangers associated with defeating such systems; and

   (ii) adherence to all occupancy restrictions;

(b) be able to recognize malfunction of the specific devices assigned to them that may result in high field strength exposures;

(c) be aware of, and trained in, normal safe operating practices and the procedures to be followed in the event of malfunction of the devices, or in an emergency situation; and

(d) use the protective equipment provided, as necessary.

7.5 Responsibilities of manufacturers, vendors and suppliers

Equipment manufacturers are responsible for making equipment that conforms to the appropriate standards within the country and for providing information on the hazards and any precautions to be taken when operating and servicing power frequency equipment. Vendors and suppliers of power frequency equipment are responsible for requesting such information from the manufacturer and providing it to the owner or employer.

7.6 Surveillance and monitoring of the workplace

The objective of field strength surveillance is to determine from records whether the equipment or installation complies with recommended standards of performance and personnel exposure (i.e. whether or not excessive exposure has occurred), to delineate boundary areas requiring shielding and to identify controlled and uncontrolled areas in the workplace.

7.6.1 Survey procedures and data collection

When conducting field strength surveys, the following procedure is recommended:

(a) field strength surveys should be carried out by a competent person, preferably the health physicist;

(b) surveys should be conducted in the following situations:

(i) before routine operations begin, for all new installations capable of producing electric or magnetic fields exceeding the limits recommended;

(ii) when any malfunction is suspected that may affect field strength levels;

(iii) following any repairs or changes in working conditions, protective shielding and barriers that may affect the exposure levels, to ensure that the levels do not exceed the recommended exposure limits;

(iv) at two-year intervals at installations capable of exposing personnel in excess of the recommended exposure limits;

(c) survey instruments must be selected to match the operating conditions and should be calibrated at least once every year. Their calibration should also be checked against another instrument before carrying out a field strength survey;

(d) during a survey, a complete record should be kept of the field parameters (electric field strength, magnetic field strength) at each work site in order to make a realistic evaluation of potential personnel exposure;

(e) during the inspection of any device or installation, all safety interlocks, "ON-OFF" control switches, and required warning signs, labels and tags should be examined.

Records should be kept of all formal field strength survey measurements and their evaluation, including the date that measurements were made, the number and types of device in the area surveyed, the geometrical coordinates of the site surveyed, the field strengths measured, and the model and serial number of the field strength survey instruments used. Such records should also include a review of all known incidents and their attributed causes.

7.6.2 Assessment and interpretation of occupational exposure data

In assessing and interpreting the occupational exposure data, the following steps are necessary:

PROTECTION FROM POWER FREQUENCY ELECTRIC AND MAGNETIC FIELDS

(a) the results of monitored workers should be evaluated in terms of the available exposure limits;

(b) special considerations may be required in the case of pregnant women. Because of the possible greater susceptibility of the embryo or foetus to induced current densities, pregnant women should not be exposed to levels exceeding non-occupational limits;

(c) special considerations are required in the case of persons wearing medical implants (especially cardiac pacemakers). Medical implants for workers who are occupationally exposed to electric or magnetic fields should be selected and implanted with regard to these problems;

(d) workers who were subject to exposures significantly higher than the recommended limits should be referred to the occupational physician for special medical surveillance. High exposure alone does not necessitate the removal of the worker from work.

7.7 Control of occupational exposure

Responsibility for protection of workers and all other persons against the potentially adverse effects of exposure to power frequency electric and magnetic fields should be clearly assigned to a department, agency, committee or individual, as indicated in the previous sections of this chapter. Such responsibility should include:

(a) the development and adoption of exposure limits, area control procedures and the implementation of a compliance programme;

(b) the evaluation of protection against emissions from existing or newly manufactured equipment and installations. The development of an emission standard for the equipment may be considered, so as to avoid the dependence for safety on instructions and labels alone. Safe design of equipment should be required so that the need for use of additional protective devices is reduced to an unavoidable minimum;

(c) the development of standardized measurement procedures and survey techniques. Survey reports should be retained and should include details of the exposure conditions and, in cases where exposure limits are exceeded, indications of ways and means for reducing levels.

7.7.1 Controlled areas

Areas where surveys indicate that exposure limits can be exceeded should be designated as areas of controlled access requiring the development of safety proce-

CONTROL OF EXPOSURE TO 50 AND 60 Hz FIELDS

dures to ensure that no one can be exposed to more than is allowed by the appropriate exposure standard. Access should be restricted to workers who understand the possible consequences of overexposure, who have a need to enter the area and for whom adequate supervisory controls are provided.

There are different means of controlling exposure in an area, e.g. by keeping the exposure levels down or remaining in the area for as short a time as possible to complete the task. Thus controlled areas should have warning signs that indicate the exposure level and the maximum time that should be spent in the area.

As with any potential health hazard, needless exposure should be avoided and protective measures must always be optimized for the control of health hazards. Exposure limits for protection of the worker and the general public are established to indicate levels above which a health risk is recognized and below which current scientific knowledge has not established a health hazard.

7.7.2 Protective measures for electric field coupling

Protection from direct electric field exposure can be achieved relatively easily with the use of shielding. At ELF frequencies, virtually any conducting surface will provide substantial electric field shielding. One practical approach for personnel working in high field strength regions is to provide them with clothes that are electrically conductive. This practice is used commonly in the electric utility industry by lineworkers who work on high-voltage transmission lines using "bare-hand" techniques. Deno and Silva (1984) used such a fabric to make a conductive vest for this research effort. Another method to obtain protection from electric field exposure is to limit the access of individuals to areas where electric field strengths are high.

Personnel protection from the effects of indirect coupling to ELF fields can often be obtained by appropriate grounding practices. For example, humans who are resonably well grounded may be exposed to possibly significant ELF currents if they touch a truck or bus that is located in an ELF electric field. The solution, in many cases, can consist of ensuring that the bus or truck is electrically grounded and, perhaps, equipping with insulating footwear individuals who might be exposed. The use of insulated gloves may be useful when ungrounded objects must be handled. Caution dictates the use of protective devices (e.g. suits, gloves and insulation) in all fields exceeding 15 kV/m.

Other hazards due to sparks from high voltages at large metallic structures may arise particularly in connection with the use of explosives or flammable substances. If proper shielding is not possible, the field generators must be switched off during such use.

In general, modern protective measures, when properly applied, should ensure the safe operation of electric field sources.

7.7.3 Protective measures for magnetic field coupling

There is no practical economic way to shield against direct ELF magnetic field exposure. Thus, the only protective method, if such is deemed necessary, is to limit exposure, either by limiting access of personnel to areas where magnetic flux densities exceed whatever safety standard is set or by limiting magnetic flux densities in areas where humans could be exposed.

As in the case of electric fields, ELF field exposure resulting from indirect modes of coupling to magnetic fields can be limited in many cases by appropriate grounding procedures. Consider, for example, magnetic coupling to a barbed-wire fence. Barbed-wire is commonly strung on wooden posts to make fences. Thus, an individual strand of wire may be reasonably well insulated from the ground over most of its length. If this strand is grounded at some point, for example at a metal gate, and a person touches it, a conducting loop is formed. The loop consists of that portion of the fence between the gate and the person, the person's body, the ground and the gate. An electromotive force will be induced in this loop if an ELF magnetic field is present, and currents will flow as a result.

It is easily shown, for environmental sources of ELF magnetic fields, that rather large loops are required for the generation of significant voltages (i.e. > 10 V). This fact suggests that mitigation can be obtained, if necessary, by grounding a fence at multiple intervals along its length to break any large conducting loops into a number of smaller loops.

7.7.4 Siting and installation

To assess the protective measures needed for locations in which electric or magnetic devices are routinely used, it is necessary to determine the occupancy and ownership of the surroundings. When conducting surveys around these devices, it should be noted that:

(a) exposure in uncontrolled areas should not be allowed to exceed the general public limits;

(b) field strength levels should be known and periodically checked as they may change in controlled areas where restricted occupancy is allowed. These areas should be designated and the maximum occupancy time posted;

(c) the immediate vicinity of unmanned, high-field sources should be fenced off and warning signs posted;

(d) electric and magnetic field sources and devices should be positioned as far away as is necessary from areas normally occupied by other persons not occupationally exposed;

(e) there should be no unnecessary metallic objects near strong field producing devices. The presence of such objects may result in high-intensity fields in some locations, and may increase the possibility of shocks and burns;

CONTROL OF EXPOSURE TO 50 AND 60 Hz FIELDS

(f) shielding or screening should be achieved by use of appropriate materials. When screening sources with enclosures not designed for occupancy, reflective materials such as sheet metal may be used.

7.7.5 Warning signs and labelling

Warning signs should be used to indicate the nature and degree of hazard associated with a given device or location. The size of the sign should be such that it is clearly distinguishable under the appropriate working conditions, being either illuminated or employing reflective materials as necessary.

It is suggested that the CAUTION sign be **black** on a contrasting colour such as a **yellow** background, the WARNING sign **black** on, for example, an **orange** background, and the DANGER sign **red** on a **white** background. Where personnel are employed at night or in darkened areas, illuminated signs should be provided.

The CAUTION sign is not generally used for area control but should be placed on devices to indicate that they produce electric or magnetic fields. Both the WARNING and DANGER signs are used for identifying hazardous devices and for area demarcation.

When the signs are used to designate an exposure area or zone, the following recommendations apply:

(a) the WARNING sign should be placed at the entrance of any zone within which a survey has shown that electric or magnetic field strength values exceed the occupational limits. The WARNING sign indicates that **limited occupancy** is allowed within its boundaries. It should be located wherever it is necessary to indicate a limit on the occupancy time. In such cases, the WARNING sign should be accompanied by words such as "Warning: High electric/magnetic field strengths – Maximum occupancy time (t) minutes per hour", where (t) is calculated according to the field exposure limits recommended;

(b) the DANGER sign should be placed at the entrance of any zone where a survey has been conducted and where electric or magnetic field strengths occur in excess of five times the continuous occupational exposure limit for electric fields and 20 times the continuous occupational exposure limit for magnetic fields. The DANGER sign indicates a **denied occupancy** zone and should be used in conjunction with barriers that are intended to prevent human access. When used for area demarcation, the DANGER sign should be accompanied by words such as "Danger: High electric/magnetic field strengths – Do not enter";

(c) the signs for the areas or zones should be clearly distinguishable at some distance.

When the signs are used for labelling a device, the following recommendations apply:

(a) the WARNING sign should be applied to any device under development or in use for any industrial, scientific or medical purpose, if the device produces exposure

55

levels that exceed the recommended limits. The WARNING sign should be applied to a device if misuse or failure could cause injury;

(b) the DANGER sign should be applied to any device under development or in use for any industrial, scientific or medical purpose, if it produces exposure levels in excess of five times the continuous occupational exposure limit for electric fields and 20 times the continuous occupational exposure limit for magnetic fields.

Appendix A – Biological effects of ELF electric and magnetic fields

Biological effects and potential health consequences of exposure to sinusoidal electric and magnetic fields at power frequencies have been the subject of several thousand scientific research papers. This appendix reviews the principal findings and conclusions relevant to the assessment of possible health effects of exposure at power frequencies. However, some selected data on effects of exposure at other ELF frequencies, and to other than sinusoidal wave-forms, will also be discussed.

Assessment of health effects of exposure can be based on two complementary approaches. One consists of a "black-box", "short-cut" approach where purely empirical observations are collected in an attempt to relate external unperturbed fields to changes in biological endpoints deemed important for maintaining normal structure and function (i.e. health) of the exposed organism. This may be refined by relating external to internal fields, followed by an attempt to establish a relationship between the magnitude of the internal fields and the extent of observed biological changes. Estimates of such a relationship can form the basis for preliminary guidelines on exposure limits. The second approach consists of developing a predictive theory derived from a mechanistic understanding of interactions of electric (E) and magnetic (H) fields with living systems. Such understanding is developed from the available knowledge in the fields of biophysics and bioelectrochemistry. At present, the available understanding and knowledge are incomplete, and so only limited use of the second approach is possible.

The effects of electric and magnetic fields are sometimes difficult to dissociate. Time-varying H-fields induce electric fields and currents in conducting objects. Thus many effects of external ELF magnetic fields are due to internally induced electric currents and E-fields. Phenomena associated with external E-field strengths occurring in practice, and which may possibly occur, are described in section A1. Phenomena associated with H-fields encountered in practice in the environment are described in section A2. Thus, both sections A1 and A2 have to be read and considered together, and in conjunction with both sections 4.2 and 4.4 in the main text on mechanisms of field interactions.

PROTECTION FROM POWER FREQUENCY ELECTRIC AND MAGNETIC FIELDS

A1 Biological effects of electric fields

A1.1 Introduction

An in-depth analysis and conclusions representing an international consensus can be found in the UNEP/WHO/IRPA (1984) document, *Extremely low frequency (ELF) fields, Environmental Health Criteria 35*. The conclusions of this document were reaffirmed by another international group of experts convened by the WHO Regional Office for Europe (WHO-EURO, 1989). Extensive bibliographies have been prepared by Bridges (1975), Sheppard and Eisenbud (1977), Kornberg and Sagan (1979), Kavet (1979) and Carstensen (1987). Additional information can be found in symposia proceedings (Grandolfo et al., 1985, Chiabrera et al., 1985) and monographs (Blank and Findl, 1987, Polk and Postow, 1986).

A1.2 Animal experiments

It is impossible to discuss the large number of animal experiments performed to date. Only a brief summary is given. The overwhelming majority of experiments were designed using the "black-box" approach and relate the effects to unperturbed incident field strengths. Only recently have dosimetric approaches been developed to determine the field strength at the surface of the body and current densities in various parts of the body in relation to unperturbed strength of the incident field (see Anderson and Phillips, 1985; Kaune and Forsythe 1985, 1988). This section will be devoted only to recent studies which can be interpreted using such approaches. Few studies have been performed at low *E*-field strengths. In most studies, exposure levels cluster around 30 kV/m and 100 kV/m.

Although the majority of experiments were performed using rats and mice, a wide variety of species was used, including rabbits, cats, dogs, swine and primates.

A1.2.1 Behaviour

Behavioural studies indicate that threshold levels for changes were related to the detection of the presence of the field and consist of alterations in locomotor activity, and aversion or preference for the area in which the field is present. An important mechanism for field detection is hair vibration due to field forces exerted on the charge at the surface of the hair. The phenomenon was studied at 50 and 100 Hz (Cabanes and Gary, 1981) and 60 and 120 Hz (Deno and Zaffanella, 1982). Density of hair on skin and enhancement of fields on body surfaces play a role. Furry animals, rats and mice, may detect the field at unperturbed strengths of 1.2-4.0 kV/m.

Subtle effects on behaviour were reported at levels below perception. Up to 100 kV/m behavioural changes in animals are related to perception and do not affect performance in any significant manner.

A1.2.2 Nervous system

The nervous system, because of its exquisite organization and complexity, and the electrical nature of nerve impulse propagation and synaptic relations, seems to be a likely target for electromagnetic interference.

58

APPENDIX A

Studies on neurotransmitters yielded conflicting results. Various shifts were reported, mostly at unperturbed field strengths above 5 kV/m, tens of kV/m or about 100 kV/m. Reported changes were slight. The same can be said about studies on electrical activity of the brain (EEG). Levels of neuroendocrine substances and melatonin synthesis in the pineal gland may be affected by exposure of at least three weeks' duration to levels of 40 and 1.5 kV/m (Wilson et al., 1981; Anderson et al., 1982).

A1.2.3 Other endpoints

Changes in corticosterone and testosterone levels were reported in blood and urine following exposures to 50 or 60 Hz E-fields above 15 kV/m. No significant changes were found in serum. Reports on effects of exposure on blood cell counts show conflicting results.

The weight of evidence indicates that no significant adverse effects have been demonstrated in the nervous, neuroendocrine, cardiovascular, blood-forming and immune systems and in the metabolism of laboratory animals exposed for prolonged periods to unperturbed sinusoidal 50 or 60 Hz E-fields at strengths of the order of tens of kV/m, about 30 kV/m for larger animals and up to 100 kV/m for small laboratory animals.

A six-year, three-generational study of miniature swine exposed to 60 Hz vertical sinusoidal E-fields of unperturbed field strength at ground level of 30 kV/m was conducted by a large team of investigators at Battelle Pacific Northwest Laboratories (Anderson and Phillips, 1985). A large battery of tests was used to investigate many biological endpoints. Detailed and extensive examinations did not reveal any statistically significant and consistent differences between the exposed and sham-exposed populations in the following: indices of growth; haematological, biochemical and immunological studies of peripheral blood; peripheral nerve function; endocrinology; and behaviour. It should be stressed that cytogenetic studies of peripheral blood lymphocytes also did not reveal any differences between the exposed and sham-exposed group, nor were any constitutive karyotypic aberrations found.

A1.2.4 Reproduction and development

In the Battelle study quoted above, reproduction and development was assessed twice in both the F_0 and F_1 generations (Sikov et al., 1985). Compared to the sham-exposed controls, the exposed groups showed the following:

– fewer prenatal deaths and consequently more live foetuses in the first F_0 mating (four months' exposure);

– lower body weights and increased incidence of malformations in the second F_0 mating (22 months' exposure);

– impaired mating performance of F_1 females during the first mating (prenatal and 18 months' postnatal exposure);

– lower body weight and increased incidence of malformations in offspring from the first mating of F_1 females (prenatal and 18 months' postnatal exposure).

No statistically significant differences were found in foetal and birth body weights and incidence of malformations in offspring from the first F_0 mating (four months' exposure) and from the second F_1 mating (prenatal and 32 months' postnatal exposure).

PROTECTION FROM POWER FREQUENCY ELECTRIC AND MAGNETIC FIELDS

These results are difficult to interpret, partly because of the outbreak of swine flu during the study.

Contradictory results were obtained by experiments with mice and rats. Experiments with rats exposed to 60 kV/m 60 Hz fields performed by Sikov et al. (1985) did not reveal any effects of exposure on intrauterine development, and may indicate that a six-day exposure of males or females before mating and during mating does not affect its outcome.

A1.3 In vitro *studies*

A large body of evidence indicates that the cell membrane is most likely the primary site of interaction of electromagnetic fields with cells (see Blank and Findl, 1987, and Chiabrera et al., 1985, for extensive reviews and lists of references). Cellular and *in vitro* studies offer the best opportunity to explore mechanistic hypotheses and to gain fundamental knowledge about bioelectrochemistry and biophysics.

Electrical signals play an important role in information transfer. A classic example is nerve cells. A transient increase in the electric potential across the cell membrane above the resting potential is generated and propagated on the cell membrane in the form of what is called an action potential, i.e. a short significant depolarization. Such signals are spontaneously generated in neuron pacemaker cells of the marine mollusk *Aplysia*. The frequency spectrum of this activity (firing of action potentials) encompasses primarily the 0-1,000 Hz range. Wachtel (1979) studied firing pattern changes and transmembrane currents induced by ELF fields in *Aplysia* pacemaker neurons. The main conclusions of this study are that the threshold for changes depends on the direction of the field and is strongly frequency dependent. For the worst case of neurons firing at the rate of 50 or 60 Hz and the effective orientation of the field with respect of the neuron, the threshold current density for perturbation may be as low as 0.01 A/m^2 (Wachtel, 1979).

Another interesting finding is the frequency and amplitude dependence of calcium exchanges from chick or cat brain. This observation led to the development of what is called an amplitude frequency window effect. Later studies by Blackman and associates (Blackman, 1985) provided additional evidence for the existence of amplitude and frequency windows for ELF-induced calcium efflux. Although calcium ions play a significant and well-established role in regulation of cell function, the functional consequences of ELF field induced modulation of calcium ion shifts remains unclear. Further studies by Albert et al. (1987) have not been able to reproduce the *in vitro* findings on calcium efflux.

In summary, cellular and *in vitro* studies significantly contribute to the proposals for interaction mechanisms of ELF fields with living systems. Other cellular studies indicate that field strength (amplitude) and frequency dependence of effects is complex. Therefore extrapolation of conclusions from findings at a given ELF frequency to another or from one biological model to another requires the utmost caution.

A2 Biological effects of magnetic fields

A2.1 Possible field interactions

The general introductory statements made at the beginning of this appendix, and in section A1, are equally relevant to this section. Thus the reader should consider both sections jointly, and also together with the two sections 4.2 and 4.4 of the main text.

APPENDIX A

Thinking in terms of unperturbed external field strengths, it is well to remember that to a rough approximation at 50 or 60 Hz, the ratio of the internal to the external E-field is about $4 \cdot 10^{-6}$ in living systems. ELF magnetic fields induce E-fields in accordance with Faraday's law. For a given inductive loop of a radius of 0.1 m in a biological object at 60 Hz, 0.5 mT is required to produce a current density of 1 mA/m^2. The E-field induced along that circular path is about 10 mV/m. To obtain the same internal field by an external sinusoidal 60 Hz electric field, an electric field strength of 5 kV/m at 60 Hz is required. This example illustrates best the difference in the relationship between external electric and magnetic field intensities and quantities characterizing internal fields.

A2.2 Selected in vitro studies

There is a large body of evidence that pulsed magnetic fields with complex wave-forms affect a wide variety of cell functions. Such wave-forms are used in clinical practice for treatment of bone non-union, and were reported to affect bone growth and remodelling *in vivo*. Of possible health significance are effects of exposure to such pulsed H-fields on gene transcription and translation, as well as a series of *in vitro* studies on effects of exposure to such fields on chromosomes of salivary glands of larvae of *Sciara coprophila* and *Drosophila melanogaster* (Goodman et al., 1987 a, 1987 b).

Other *in vitro* studies, mostly at 50 and 60 Hz frequencies and in sinusoidal wave-forms, are summarized in UNEP/WHO/IRPA (1987). Many subtle effects were reported, but evident health implications cannot be derived from these observations.

In fact, experiments on effects of magnetic fields indicate that the induced electric field strength quantity is the important characteristic (Greene et al., 1991).

A2.3 Nerve and muscle cells, nervous and cardiovascular systems

These studies are reviewed extensively in UNEP/WHO/IRPA (1987). Data relevant to health assessment are discussed by Bernhardt (1979, 1985) and Tenforde and Budinger (1986). From experimental data, mathematical modelling and mechanistic considerations, ranges of current densities induced by ELF electric and magnetic fields can be correlated with bioeffects.

With regard to health risk assessment, several ranges of current densities may be considered. Table 6 summarized induced current density ranges at 50 or 60 Hz for producing biological effects (UNEP/WHO/IRPA, 1987) and values of the magnetic flux density for approximately inducing these current densities in peripheral regions of the heart and of the brain at 50 or 60 Hz.

A2.4 Genetic effects, reproduction and development

Sinusoidal wave-forms have not produced any effects in laboratory *in vitro* and animal studies. Particularly convincing evidence comes from the New York State Power Lines Project (Ahlbom et al., 1987) in which the combined effects of E- and H-fields were investigated. No effects were found on chromosome structure and number, and sister chromatid exchanges in somatic cells. Negative dominant lethal test results in mice indicate absence of effects on germ cells. No effects on foetal development were found following exposures of pregnant mice. Ahlbom et al. (1987) also state that no evidence of effects on human intrauterine development is available, nor are such effects likely at field strengths associated with power lines.

Appendix B – National exposure standards[1]

Although the use of electrical appliances has been widespread for several decades, and the exposure to electric fields has continuously increased, protection problems had been largely ignored until the end of the 1960s, when the early studies of medical teams (Asanova and Rakov, 1966; Sazanova, 1967) were published, reporting a number of subjective complaints by high-voltage substation workers.

The findings of early investigations stimulated a large number of studies on a variety of possible effects of electric fields on biological systems. The results are still controversial: they show that a biological response, if any, can be observed at levels of the electric field which can be practically experienced only in the vicinity of high-voltage transmission lines. Therefore, standards which have been issued or are being evaluated deal almost exclusively with power lines and related facilities, such as high-voltage substations.

A general overview of the present status of regulations, and of problems associated with electric and magnetic fields from power transmission systems in industrialized countries, is given by the results of an international survey performed by the Study Committee 36 of the International Conference on Large High-Voltage Electric Systems (CIGRE, 1986).

The aim of the survey was to obtain general information about:

– experience gained in developed countries with large high-voltage networks;

– public attitudes;

– knowledge of fields generated by power lines and their physical effects;

– research on field effects either in progress or planned;

– legal matters such as regulations, disputes and legal actions taken by opposers.

For this purpose, a questionnaire was distributed in all the member countries of the Study Committee. These countries are 21 in all, with transmission lines operating at various voltage levels, as indicated in table B1, where the respective route lengths are also listed.

The results of the survey show that, even in developed countries with extended electric networks, the levels of fields generated by the lines and the related problems are not completely appreciated. For example, only 17 countries have the capability to make calculations of electric fields, and not all have access to instruments for their experimental measurement.

---

[1] The names of the countries used in this book are consistent with the dates of the relevant texts and standards cited.

APPENDIX B

Table B1. Operation voltage and transmission route length of networks in countries participating in the CIGRE survey

Country	Voltage (kV)	Length (km)	Country	Voltage (kV)	Length (km)
Australia	500	1 420	Netherlands	400	574
	330	4 800	Norway	420	1 140
	330	3 640		300	3 570
Belgium	400	802	Poland	750	114
Brazil	750	570		400	2 140
	500	9 260	South Africa	400	8 926
	345	6 800		275	5 965
Canada	750	9 600	Spain	380	8 496
	500	9 060	Sweden	400	1 000
	354	7 600	Switzerland	400	1 000
Czechoslovakia	400	4 500	United Kingdom	275	1 693
Denmark	400	365	United States	765	3 100
Germany, Fed. Rep.	380	10 250		500	32 000
Finland	400	3 200		345	45 000
France	400	9 013	USSR	1 150	1 905
Italy	420	5 300		750	3 800
Japan	500	3 456		500	34 000
	275	6 776		330	27 000

As far as exposure standards are concerned, almost half (ten out of 21) of the countries have no regulations or guidelines giving limit values for the electric field under lines. For the others, five out of 11 have limits imposed or recommended by some public agency or department, whereas in the remaining six countries some protection is assured at the time of the design of a new line, by ensuring that the project conforms to design principles established by the utilities themselves.

Owing to the relatively scarce knowledge about biological effects, and to the different degree of appreciation of the health hazard, the reasons given for limits may be different. In some standards, limits aim at reducing the discharge current from large objects, and in others at avoiding microshocks and similar unpleasant effects; in still others, they simply tend to restrict long-term exposure to high fields.

In general, limits which are currently adopted in a given country or region are such as to pose no significant constraint on the existing design practice, taking into account the geographic, economic and demographic characteristics of the areas involved (CIGRE, 1986). The problem of possible health effects of 50/60 Hz magnetic fields has so far received much less attention than electric fields. A thorough review of the scientific literature on the biological effects of magnetic fields is given in two Environmental Health Criteria documents developed by the IRPA/INIRC in cooperation with the World Health Organization (UNEP/WHO/IRPA, 1984, 1987).

The current knowledge of biological effects seems indeed inadequate to confirm or disprove any possible health hazards. Nevertheless, in recent years several authors have reported possible correlations between cancer and prolonged presence near electric wiring characterized by high current intensities, both in houses and in workplaces. Some of the authors suggested that the magnetic field, rather than the electric field, may act as a causal or at least as a promoting factor. Although these findings have been contradicted by the results of other authors, which show no apparent difference between exposed and control groups, the above-mentioned papers have raised interest in the health effects of AC magnetic fields, and a number of well-designed surveys are being planned.

At the time of the survey, only the USSR had issued an official regulation limiting exposures to 50 Hz magnetic fields. Some agencies, however, are examining the possibility of establishing standards on ELF magnetic fields. In two countries, namely the Federal Republic of Germany and the United Kingdom, a proposal for limits has been published.

In the following, standards are examined in more detail. They are limited to those issued by national authorities or public agencies independent of the electric utilities (Grandolfo and Vecchia, 1989; Repacholi, 1992).

Australia

At the 108th session of the National Health and Medical Research Council, the interim guidelines on limits of exposure to 50/60 Hz electric and magnetic fields were approved. These guidelines (NH & MRC, 1989) are exactly the same as the international guidelines published by IRPA in 1990. The NH & MRC guidelines have received widespread acceptance in each of the states.

Czechoslovakia

Some information about present regulations in Czechoslovakia is given by Kabrhel (1985).

For presently operated 400 kV power lines, a Right-of-Way (RoW) area is required, up to a distance of 25 m from the outer phase. Provided safe conditions are ensured, agriculture is allowed within the RoW area. Building or living in this area is prohibited. Cars and agricultural machines are not allowed to stop below the line.

For workers, a guide published concerning workers in 400 kV substations requires new substations to be designed so as to limit the electric field to 15 kV/m. Under these circumstances, no further precautions are used.

Federal Republic of Germany

In the Federal Republic of Germany a draft of a national standard for the protection of persons against hazards from electromagnetic fields in the frequency range from 0 Hz to 3,000 GHz was published in 1986, and submitted to the public for objections and suggestions. The part of the standard concerning the frequency range between 0 Hz and 30 kHz has been approved (FRG 1989).

For power frequency electric fields the proposed standard sets a limit for the peak amplitude, which is 1.5 times the limit for the rms field strength; their values are 30 and 20 kV/m, respectively.

The definition of the limit value is based on the unperturbed homogeneous field. In the case of an inhomogeneous electric field, an equivalent homogeneous field must be calculated through measurement procedures that allow a comparison with the total current dispersed through the human body in the unperturbed field.

Exposure to electric fields below the limits is allowed for unlimited time. For short-time exposures (up to 2 hours per working day), exposure to field strengths up to 1.5 times the limit value is allowed. It is recognized that some people may feel oppressed by electric fields of intensity of 2 to 10 kV/m, or may experience unpleasant electrical discharges when touching charged objects or other people in an electric field, but no scientific data exist at present which allow these sensations to be related to health effects.

The standard sets limits also for power-frequency magnetic fields. The proposed standard sets a limit for the peak amplitude, which is 1.5 times the limit for the rms strength; their values are equal to 6,000 and 4,000 A/m, respectively.

The definition of the limit value is based on the case of a homogeneous field. In the case of an inhomogeneous magnetic field, an average is to be calculated over a circular area of 100 cm^2.

The above limits are intended for unlimited exposures of the whole body. For prolonged exposures of the extremities (hands, arms, feet and legs), a value up to 10 times the limit is allowed. For short times (up to 5 minutes per hour), exposure to fields of strength up to 1.5 times the limit value is permitted. The scientific basis of the current density concept which is partially used is given by Bernhardt (1985). Effects of currents generated in tissues have been considered, whereas effects which are only related to the perception of the field, but do not constitute any health hazard, such as magnetophosphenes, have been disregarded.

Japan

In Japan all electrical equipment is subject to the Technical Standards of Electrical Facilities Ordinance of the Ministry of International Trade and Industry, issued in 1973. This ordinance (Yasui, 1985) includes some specifications aimed at the safety and protection of the public.

No legal standards exist for workers at substations. As reported by Yasui (1985), possible measures to protect personnel from unpleasant effects of transient discharge due to static induction were examined by a study committee composed of members from the Central Research Institute of the Electric Power Industry, from the electric power companies and from manufacturers, which established design principles for substations. For example, 275 kV substations are designed in such a way as to limit the field strength above the ground to less than 7 kV/m.

Poland

In Poland an official standard was enacted in January 1980 to protect people and the environment against non-ionizing electromagnetic fields. In the frame of a general regulation for fields of frequencies up to 300 GHz, limits are also established for the single frequency of 50 Hz.

As a general rule, the maximum power-frequency electric field strength from power lines should not exceed 10 kV/m.

To avoid exposure to fields higher than this limit, a protection zone is defined as the area along the line where the electric field exceeds 10 kV/m. In this zone, the presence of anyone other than personnel working on the line is forbidden. A second protection zone is also defined as the area where the field strength ranges between 1 and 10 kV/m. In this latter, only the temporary presence of people related to farming, touring and recreation is allowed.

According to the response of Polish authorities to the CIGRE questionnaire (Pilatowicz, 1985), only the second protection zone is of practical importance, since the field levels under power lines do not actually reach 10 kV/m.

For permanent stay, as well as for the construction of residential houses and buildings requiring special protection against field effects, such as hospitals, schools and nurseries, the maximum admissible field strength amounts to 1 kV/m.

For workers, a limit of 15 kV/m is established for the electric field strength in high-voltage substations. In special circumstances, an exposure to fields up to 20 kV/m is allowed during maintenance operations; in this case, permanent screens must be used.

United Kingdom

The National Radiological Protection Board (NRPB, 1989) issued guidance on limits of human exposure to electromagnetic fields in the frequency range 0-3,000 GHz. These limits were recommended following the release for comment of two consultative documents in 1982 and 1986. Limits of exposure to 50 Hz fields are recommended to be the same for workers and the general public. The NRPB recommends that continuous exposure to electric fields be limited to 12.28 kV/m and magnetic fields to 2 mT (1,630 A/m). In addition, continuous induced current in any arm, hand, leg, ankle or foot should not exceed 1.03 mA.

The basis for the restrictions in the United Kingdom is to protect against the possibilities of electric shocks and burns. The advised restriction on induced current is compared to the allowable leakage currents for domestic electric appliances of between 0.25 and 5.0 mA recommended by the British Standards Institute in 1985. The NRPB considers that the safety factors incorporated in the above limits are sufficient for both the general public and occupational exposure and cannot find any scientific justification or need for further safety factors for general public exposure (unlike the separate limits in the IRPA 1990 guidelines).

United States

In the United States regulations widely differ from one state to another. A review of the existing standards has been performed by Shah (1979); more recent data have been given by Banks (1986), Zaffanella (1985) and Repacholi (1992).

For the design criteria of new lines, all the states have adopted the National Electric Safety Code (NESC), or some modification of it, *for practical safeguarding of persons from hazards arising from the installation, operation and maintenance of overhead supply and communication lines and their associated equipment.*

The code requires the lines to be designed so as to limit the discharge current to 5 mA if the largest truck, vehicle or equipment under the line were short-circuited to the ground.

The concept of Right-of-Way (RoW) is adopted by electric companies throughout the country. No house, and no full-time activity, is allowed in the RoW. The width of RoWs depends on the operation voltage of the line and the policy of the utility. Typical values are given in table B2, along with typical values of the maximum electric field measured under the lines.

Table B2. Typical values of maximum electric field and Right-of-Way widths in the United States

Voltage (kV)	Max. field strength (kV/m)	RoW width (m)
345	5	45
500	8	53
765	10	76

Source: Zaffanella, 1985.

Although most states require preliminary evaluation of public safety and comfort along the area traversed by a proposed line, only a few have recommended guidelines for maximum permissible electric fields.

As can be seen from table B3, the standards are not consistent, since each of the states has different limits inside or at the boundary of the RoW. The limits within the RoW are intended to avoid currents above let-go threshold for people touching vehicles at road crossings. Those at the edge of RoW were set to minimize the risk of possible health effects due to long-term exposures.

The existing limits are aimed at population protection. No fixed rules exist for workers in substations. This is due to the fact that, because of station automation and conservative design, the time spent by workers in high electric fields is small.

APPENDIX B

Table B3. **Typical values of electric strength under power lines recommended in the United States**

State	Max. field strength (kV/m)	Comments
Minnesota	8	Within RoW
Montana	7	At road crossings
	1	At edge of RoW
New Jersey	3	At edge of RoW
New York	11.8	Within RoW
	11	At private roads
	7	At public roads
	1.6	At edge of RoW
North Dakota	8	Within RoW
Oregon	9	Within RoW

Source: Zaffanella, 1985.

USSR

The USSR was the first country to issue official exposure standards for electric fields generated by transmission lines, as a consequence of the concern raised by the already mentioned findings of Asanova and Rakov (1966) and of Sazanova (1967). A first regulation document (USSR, 1970) regarding workers employed at AC substations and transmission lines operating at 400, 500 and 750 kV was approved by the USSR Ministry of Health on 29 October 1970, to be enforced in 1971. Five years later, the standard was replaced by a new regulation (USSR, 1975), applying to workers in substations or on transmission lines operating at 400 kV and above, so as to include people working on the new Extremely High-Voltage (EHV) systems at 1,150 kV. The regulation was enforced on 1 January 1977, to last five years. Its rationale is given by Lyskov et al. (1975).

Since 1984, occupational exposure to 50 Hz electric fields is regulated by the State Safety Standard 12.1.002-84 of Safety Standard System *Industrial Frequency Electric Fields. Permissible strength levels and control requirements at the workplaces.* In this document the maximum permissible level is fixed as 25 kV/m and, if the electric field strength is higher, there is a special need for the use of shielding or screening suits. If the electric field level is higher than 20-25 kV/m, the permissible working time is not longer than 10 minutes. Work in areas where the electric field strength is lower than 5 kV/m inclusive is allowed for the whole working day.

The permissible working time for the strength range 5-20 kV/m inclusive is determined by the formula:

$T = 50/E - 2$, where
T = permissible time (in hours) and E = electric field strength (in kV/m).

A permissible time may comprise one or several periods over the working day.

The regulation of population exposure to industrial frequency electric fields is realized according to the Ministry of Health document *Sanitary norms and regulations of population protection from AC industrial frequency electric fields of overhead transmission lines*, N 2971-84.

In this document maximum permissible field strengths are fixed for different environments:

PROTECTION FROM POWER FREQUENCY ELECTRIC AND MAGNETIC FIELDS

0.5 kV/m   –   for the housing areas and other residential buildings

1.0 kV/m   –   for the land around residential buildings

5 kV/m   –   outside residential areas

10 kV/m   –   in zones crossing I-IV class roads

15 kV/m   –   in unpopulated areas

20 kV/m   –   in remote areas

The Ministry for Public Health of the USSR (USSR, 1985) has issued the only official standard regulating exposure to 50 Hz magnetic fields. In the standard, a distinction is made between continuous and pulsed field. The time of exposure is limited, depending on the pulse characteristics, as shown in table B4.

The standard seems to have been developed for arc welding, which constitutes the main source of occupational exposure to pulsed magnetic fields at power frequency.

No rationale for this standard appears to have been published so far.

Table B4. **Maximum permissible levels of 50 Hz magnetic fields (A/m) in the USSR (1985)**

Duration of exposure (h)	Pulse characteristics		
	Continuous or >0.02 s; $T \leqslant 2$ s	$1s<t<60$ s $T>2$ s	0.02 s$<t<1$ s $T>2$
1.0	6 000	8 000	10 000
1.5	5 500	7 500	9 500
2.0	4 900	6 900	8 900
2.5	4 500	6 500	8 500
3.0	4 000	6 000	8 000
3.5	3 600	5 600	7 600
4.0	3 200	5 200	7 200
4.5	2 900	4 900	6 900
5.0	2 500	4 500	6 500
5.5	2 300	4 300	6 300
6.0	2 000	4 000	6 000
6.5	1 800	3 800	5 800
7.0	1 600	3 600	5 600
7.5	1 500	3 500	5 500
8.0	1 400	3 400	5 400

t = pulse width duration; T = pulse pause duration.

In 1989 the Ministry of Health issued a special document for the magnetic field (MF) under live-line maintenance regulation: *Approximate safety levels of industrial frequency alternating magnetic fields under bare-hand maintenance on 200-1,150 kV transmission lines* (N 5060.89).

According to this document the maximum permissible level of total and local magnetic field exposure is 3.2 and 5.2 kA/m, respectively, and the permissible time is four hours per working day.

Appendix C – Glossary

Alternating current (AC). An electric current varying sinusoidally in time.

Alternating electric field. The electric field produced by a sinusoidally oscillating electric charge.

Alternating voltage. A voltage varying sinusoidally in time.

Biophysical. Relating to the physical properties of biological systems, e.g. the conductivity of tissue is a biophysical quantity.

Conductivity. The scalar or matrix quantity whose product by the electric field strength is the current density.

Controlled area. An area in which the exposure of personnel to an electric or magnetic field is under the supervision of a safety officer.

Controls. Animals, tissue, etc., not subjected to the field or other experimental treatment (see also **Sham-exposed**).

Current density. The flow of electric current across a unit area, a measure of the distribution of current within the object or body tissues measured in amperes per square metre (A/m^2), or microamperes per square centimetre ($\mu A/cm^2$).

Earth. Electrical ground.

Effective field. The time-averaged electric field to which a biological system is exposed; this field is less than the unperturbed field because of mutual shielding, e.g. by animals housed as a group.

Electric field (E). Concept used to represent the force exerted on the unit charge due to the location of electric charges at various sites in a region; the high-voltage transmission line electric field is an alternating (50 or 60 Hz) field due to the sinusoidally oscillating charges located on the conducting wires of the transmission line. Electric fields are capable of performing work on other electric charges moving between points at different potential.

Electric field strength. Magnitude of the electric field, measured in volts per metre; (see **Volt per metre**). The electric field strength beneath a HV transmission line is generally measured at a fixed height above ground (usually 1 m).

ELF. Abbreviation for extremely low frequency.

Employer. The person responsible for administrative controls over given areas and employees (workers).

Exposure, partial body. Exposure of only part of the body to an electric or magnetic field.

Exposure, whole body. Exposure of the entire body to an electric or magnetic field.

Extremities. Limbs of the body including the arms, wrists, legs and ankles.

Faraday cage. Grounded cage made of conducting material used to enclose an object subjected to an electric field; the shield, usually composed of metal (e.g. copper wire), reduces the electric field strength inside the cage to nearly zero.

Field. A region of space in which certain phenomena occur, described by a scalar or vector quantity, the knowledge of which allows the effects of the field to be evaluated.

Field strength. The magnitude of a component of specified polarization of the electric or magnetic field. The term normally refers to the rms value of the electric field.

Free space. An ideal perfectly homogeneous medium that possesses relative dielectric and magnetic constants of unity, and in which there is nothing to reflect, refract or absorb energy. A perfect vacuum possesses these qualities.

Ground. Zero potential, electric earth.

Hertz (Hz). The unit of frequency for a periodic oscillation corresponding to a complete oscillation or cycle per second (cps).

High tension line. A high-voltage transmission line.

Horizontal electric field. An electric field directed parallel to the Earth's surface; in the laboratory, such fields are created by vertical plates.

HVTL. High-voltage transmission line, typically one operating at or above 345 kV.

Impedance. The physical property of a material that determines the relation between current flow and potential difference in the material; for alternating currents impedance includes the properties of resistance, capacitance and inductance.

Impedance to ground. An impedance measured between an object and earth (ground); in caged laboratory animals, this property depends on the caging materials, construction and electrical design, and the biophysical properties of the animal's footpad; for human beings, skin, clothing or shoe properties are significant.

Interlock. A device that prevents the activation of a given mechanism or process unless certain precautionary measures have been observed.

Internal electric field. Electric field strength measured or calculated for points within the body of an animal or human being exposed to an external electric field.

Limited occupancy area. An area within which the field strength is greater than that specified for unlimited occupancy but less than a set limit.

Magnetic field (H). A concept to describe the force exerted on a unit current produced by moving electrical charges, such as those in an electrical current; the transmission line magnetic field is due to the flow of current in the wires. A magnetic field exerts a force on moving electric charges, such as those in another wire-carrying current or in a moving wire (dynamo principle) always perpendicular to the direction of motion.

Neurophysiological. Relating to the function of the nervous system, e.g. peripheral nerves, brain, spinal cord, the subdivisions of those organs and their cellular components, including the nerve fibres.

Occupancy time. The time period that a worker is allowed to spend in a limited occupancy area within the averaging time of the exposure standard.

Owner. A person, organization or institution having title to administrative control over a given device producing electric or magnetic fields. The owner is the owner of the field source and is responsible to all people in the facility, not just the employees.

Responsible use. A person designated by the owner to have immediate authority for the use and operation of a given device producing electric or magnetic fields.

Right-of-Way (RoW). The provision of access to power lines for inspection and maintenance purposes; the concept varies from one country to another. It may take the form of the ownership of land over which the power lines pass, or the statutory control of access to this land, or the negotiation of agreements with the landowners. In some countries, the RoW applies to a corridor (strip of land), of a certain width, along the transmission line, in which the public access or property rights may be restricted.

rms. Root mean square, the square root of the temporal average over a period of the square of the field strength magnitude.

Safety officer. A person appointed by the owner to evaluate safety techniques and provide guidance regarding protection.

Scalar. A quantity that is completely specified by a single number against the relevant hazards.

Scaling. Relating an exposure of one animal species to another so that the effect of the electric field can be interpreted on an equal basis; because of shape and orientation dependence for both internal and external fields, human beings and animals exposed to the same unperturbed field have very different surface and internal fields.

Sham-exposed. A control experimental condition in which the animals, tissues, etc., are treated identically to the exposed objects, except that the field or other treatment is not present; distinguished from "controls" by the use of apparatus that is in all ways identical to the exposure apparatus which is not operating.

Shield. A mechanical barrier or enclosure provided for protection. The term is modified in accordance with the type of protection afforded, e.g. a magnetic shield is a shield designed to afford protection against magnetic fields.

Surface electric field. The electric field at the outer margin of the object or body; this field is influenced by the shape and configuration of a conducting body and, depending on the degree of curvature, is locally greater than the unperturbed electric field.

Teratological. Relating to abnormal anatomy, resulting in deformities, foetal death, still-birth, etc., especially in a developing or newborn organism.

Uncontrolled area. Any area that is not under the supervision of a safety officer.

Unperturbed electric field. The field that would exist at the body's location if there were no body located in the electric field. In the case of a uniform field, the field that exists far from the location of a conducting object (such as the human or animal body).

Vector. A mathematical-physical quantity that represents a vector quantity – it has magnitude and direction.

Vector quantity. Any physical quantity in which specifications involve both magnitude and direction and which obeys the parallelogram law of addition.

Vertical electric field. An electric field directed perpendicular to the Earth's surface; in the laboratory, such fields are created by horizontal electrodes.

Volt per metre (V/m). Unit of electric field strength; a field of 1 V/m is created in the centre of the midplane of two parallel plates separated by 1 m and having a potential difference of 1 V.

Bibliography

Ahlbohm, A., et al. 1987. *Panel's final report: Biological effects of power line fields*, New York State Power Lines Project. Albany, New York, New York State Department of Health.

Albert, E. N., et al.1987. "Effect of amplitude modulated 147 MHz radiofrequency radiation on calcium ion from avian brain tissue", in *Radiation Research*, 109, pp. 19-27.

Anderson, L. E.; Phillips, R. D. 1985. *Biological studies of swine exposed to 60 Hz electric fields*, Vol. 1: *Overview and summary*, EPRI report EA 4318. Palo Alto, California, Electric Power Research Institute (EPRI).

Anderson, L. E., et al. 1982. "Pineal gland response in animals exposed to 60 Hz electric fields", in *Abstract. 4th meeting BEMS, Los Angeles, California*. Frederick, Maryland, Bioelectromagnetics Society, p. 15.

Armanini, D. 1970. "Le applicazioni dei generatori di Hall" [Applications for Hall's generators], in *Elettrificazione* (Italian review of electrical applications), No. 4, Apr. 1970, pp. 184-188.

—; Brambilla, C. 1979. *Electric field measurements devices used in the 1,000 kV plant at Suvereto*, Third International Symposium on High Voltage Engineering, Milan, 28-31 August 1979. Milan, Associazione Elettrotecnica ed Elettronica Italiana.

Asanova, T. P.; Rakov, A. T. 1966. "The state of health of persons working in outdoor electric fields in 400 and 500 kV switch-yards" (in Russian), in *Gigiena Truda i Professional'nye Zabolevaniya*, 10, pp. 50-52 (in Russian). Translated by G. Knickerbocker, in *Study in the USSR of medical effects of electric power systems*, IEEE Special Publication No. 10. Piscataway, New Jersey, Institute of Electrical and Electronics Engineers (IEEE) Power Engineering Society, 1975.

Banks, R. S. 1986. *Regulations of overhead power transmission line electric and magnetic fields*, Syllabus of the International Utility Symposium on Health Effects of Electric and Magnetic Fields: Research, Communication, Regulation. Toronto, 16-19 September 1986. Toronto, Ontario Hydro.

Baroncelli, P., et al. 1986. "A health examination of railway high-voltage substation workers exposed to ELF electromagnetic fields", in *American Journal of Industrial Medicine*, 10, pp. 45-55.

Bernhardt, J. H. 1979. "The direct influence of electromagnetic fields on nerve and muscle cells of man within the frequency range of 1 Hz to 30 MHz", in *Radiation and Environmental Biophysics*, 16, pp. 309-323.

—. 1985. "Evaluation of human exposure to low frequency fields", in *AGARD Lecture Series*, Vol. 138: *The impact of proposed radiofrequency radiation standards on military operations*. Neuilly-sur-Seine, France, Advisory Group for Aerospace Research and Development, pp. 8.1-8.18.

—. 1988. "The establishment of frequency-dependent limits for electric and magnetic fields and evaluation of indirect effects", in *Radiation and Environmental Biophysics*, 27, pp. 1-27.

Blackman, C. F. 1985. "The biological influences of low-frequency sinusoidal electromagnetic signals alone and superimposed on RF carrier waves", in Chiabrera, A. et al., 1985, pp. 521-535.

Blank, K.; Findl, E. (eds.). 1987. *Mechanistic approaches to interaction of electric and electromagnetic fields with living systems*. New York and London, Plenum Press.

Bridges, J. E. 1975. *Biological effects of high voltage electric fields*, EPRI Report 381-1. Palo Alto, California, EPRI.

—; Frazier, M. J. 1979. *The effects of 60 Hz electric and magnetic fields on implanted cardiac pacemakers*, EPRI report EA 1174. Palo Alto, California, EPRI.

Broadbent, D. E., et al. 1985. "Health of workers exposed to electric fields", in *British Journal of Industrial Medicine*, 42, pp. 75-84.

Cabanes, J.; Gary, C. 1981. *Direct perception of electric fields*, CIGRE, Stockholm, paper 233-08. International Conference on Large High-Voltage Systems, Paris, CIGRE.

Calle, E. E.; Savitz, D. A. 1985. "Leukaemia in occupational groups with presumed exposure to electrical and magnetic fields", in *New England Journal of Medicine*, 313, pp. 1476-1477.

Carstensen, E. L. 1987. *Biological effects of transmission line fields*. New York, Amsterdam, London, Elsevier.

Chatterjee, I., et al. 1986. "Human body impedance and threshold currents for perception and pain for contact hazard analysis in the VLF-MF band", in *IEEE Transactions on Biomedical Engineering*, BME-33, pp. 486-494.

Chiabrera, A., et al. (eds.). 1985. *Interactions between electromagnetic fields and cells*. New York and London, Plenum Press.

CIGRE. 1986. *Electric and magnetic fields from power transmission systems. Results of an international survey*, CIGRE paper 36-09, International Conference on Large High-Voltage Electric Systems. Paris.

BIBLIOGRAPHY

Coleman, M. P.; Beral, V. 1988. "A review of epidemiological studies of health effects of living near or working with electricity generation and transmission equipment", in *International Journal of Epidemiology*, 17, pp. 1-13.

Conti, R. 1985. "Instrumentation for measurement of power frequency electromagnetic fields", in M. Grandolfo et al., 1985, pp. 187-210.

Czerski, P.; Athey, T. W. 1987. *Safety of in vivo diagnostic magnetic resonance examinations. Theoretical and clinical considerations.* Rockville, Maryland 20857: Food and Drug Administration Docket Management Branch: *Magnetic resonance diagnostic device panel recommendation and report on petitions for MR reclassification.* Docket 87 P-0214.

Dalziel, C. F. 1954. "The threshold of perception currents", in *Electrical Engineering*, 73, pp. 625-630.

—; Lee, W. R. 1968. "Re-evaluation of lethal electric currents", in *IEEE Transactions on Industry and General Applications*, IGA-4, pp. 467-476.

Dennis, J. A., et al. 1991. "Epidemiological studies of exposure to electromagnetic fields. II: Cancer", in *Journal of Radiological Protection*, 11, pp. 13-25.

Deno, D. W. 1974. "Calculating electrostatic effects of overhead transmission lines", in *IEEE Transactions on Power Apparatus and Systems*, PAS-93, pp. 1458-1471.

—. 1977. "Currents induced in the human body by high-voltage transmission line electric fields – Measurement and calculation of distribution and dose", in *IEEE Transactions on Power Apparatus Systems*, PAS-96, pp. 1517-1527.

—; Silva, M. 1984. "Method for evaluating human exposure to 60 Hz electric fields", in *IEEE Transactions on Power Apparatus and Systems*, PAS-103, pp. 1699-1706.

—; Zaffanella, L. E. 1982. "Field effects of overhead transmission lines and stations", in J. J. La Forest (ed.): *Transmission line reference book, 345 kV and above.* Pittsburg, Pennsylvania, General Electric.

Dimbylow, P. J. 1987. "Finite difference calculations of current densities in a homogeneous model of a man exposed to extremely low frequency electric fields", in *Bioelectromagnetics*, 8, pp. 355-375.

Filipov, V. 1972. "Der Einfluss von elektrischen Wechselfeldern auf den Menschen" [The effect of alternating electric current on humans], in *International Colloquium für die Verhütung von Arbeitsunfällen und Berufskrankheiten durch Elektrizität.* Cologne, Berufsgenossenschaft der Feinmechanik und Elektrotechnik, pp. 170-177.

Fole, F. F.; Dutrus, E. 1974. "Nueva aportación al estudio de los campos electromagnéticos generados por muy altas tensiones", in *Medicina y Seguridad del Trabajo*, 22, pp. 25-44.

Fotopoulos, S. S., et al. 1987. "60 Hz field effect on human neuroregulatory, immunologic, hematologic and target organ activity", in L. E. Anderson et al. (eds.): *Interaction of biological systems with static and ELF electric and magnetic fields*, 23rd Hanford Life

Sciences Symposium, Richland, Washington, October 1984. Richland, Pacific Northwest Laboratory, 1987, pp. 455-469.

Gamberale, F. 1990. "Physiological and psychological effect of exposure to ELF electric and magnetic fields on humans", in *Scandinavian Journal of Work, Environment and Health*, 16 (suppl. 1), pp. 51-54.

Gamberale, F., et al. 1989. "Acute effects of ELF electromagnetic fields: A field study of linesmen working with 400 kV powerlines", in *British Journal of Industrial Medicine*, 46, pp. 729-737.

Gandhi, O. P., et al. 1984. "Impedance method for calculation of power deposition patterns in magnetically induced hyperthermia", in *IEEE Transactions on Biomedical Engineering*, BME-31, pp. 644-651.

Goodman, R., et al. 1987a. "Transcriptional patterns in the X chromosome of *Sciara coprophila* following exposure to magnetic fields", in *Bioelectromagnetics*, 8, pp. 1-8.

Goodman, R., et al. 1987b. *Transcription and translation in* Drosophila *salivary gland cells exposed to low frequency non-ionizing radiation*, Ninth Meeting of the Bioelectromagnetics Society, Portland, Oregon. Frederick, Maryland, BEMS.

Graham, C.. et al. 1987. "A double-blind evaluation of 60 Hz field effects on human performance, physiology and subjective state", in L. E. Anderson et al. (eds.): *Interaction of biological systems with static and ELF electric and magnetic fields*. 23rd Hanford Life Sciences Symposium, Richland, Washington, October 1984. Richland, Pacific Northwest Laboratory, 1987, pp. 471-485.

Grandolfo, M.; Vecchia, P. 1989. "Existing safety standards for high-voltage transmission lines", in G. Franceschetti et al. (eds.): *Electromagnetic biointeraction. Mechanisms, safety standards, protection guides*. New York and London, Plenum Press, pp. 153-173.

Grandolfo, M., et al. (eds.). 1985. *Biological effects and dosimetry of static and ELF electromagnetic fields*. New York and London, Plenum Press.

Greene, J. J., et al. 1991. "Delineation of electric and magnetic field effects of extremely low frequency electromagnetic radiation on transcription", in *Biochemical and Biophysical Research Communications*, Vol. 174 (2), pp. 742-749.

Guy, A. W. 1985. *Hazards of VLF electromagnetic fields*, AGARD Lecture Series No. 138: *The impact of proposed radiofrequency radiation standards on military operations*, Neuilly-sur-Seine, France, AGARD, pp. 9.1-9.20.

—; Chou, C. K. 1982. *Hazard analysis: Very low frequency through medium frequency range*, Report USAFSAM 33615-78-D-0617. Brooks Air Force Base, Texas, USAF School of Aerospace Medicine, Aerospace Medical Division.

Guy, A. W., et al. 1982. "Determination of electric current distributions in animals and humans exposed to a uniform 60 Hz high-intensity electric field", in *Bioelectromagnetics*, 3, pp. 47-71.

Hamasaki, Y., et al. 1980. "OPSEF/An optical sensor for measurement of high electric field intensity", in *Electronics Letters*, Vol. 16, No. 11, 22 May, pp. 406-407.

Hauf, R. 1982. "Electric and magnetic fields at power frequencies with particular reference to 50 and 60 Hz", in M. J. Suess (ed.): *Non-ionizing radiation protection*, WHO regional Publications Office No. 10, Copenhagen, WHO, Regional Office for Europe, pp. 175-197.

IEC (International Electrotechnical Commission). 1984. *Effects of current passing through the human body*, Part 1: *General aspects*, Ch. 1: "Electrical impedance of the human body"; Ch. 2: "Effects of alternating current in the range of 15 Hz to 100 Hz"; Ch. 3: "Effects of direct current". Geneva, IEC Publication 479-1.

—. 1987. *Effects of current passing through the human body*, Part 2: *Special aspects*, Ch. 4: "Effects of alternating current with frequencies above 100 Hz"; Ch. 5: "Effects of special wave forms of current"; Ch. 6: "Effects of unidirectional single impulse currents of short duration". Geneva, IEC Publication 479-2.

IEEE (Institute of Electrical and Electronics Engineers) Committee Report. 1978. "Electric and magnetic field coupling from high voltage power transmission lines – Classification of short-term effects on people", in *IEEE Transactions on Power Apparatus and Systems* PAS-97, pp. 2243-2252.

ILO (International Labour Office). 1985. Occupational Health Services Convention, 1985 (No. 161).

IRPA (International Radiation Protection Association). 1991. *IRPA guidelines on protection against non-ionizing radiation*, by A. S. Duchêne et al. (eds.). New York, Pergamon Press.

IRPA/INIRC (International Non-Ionizing Radiation Committee). 1985. "Review of concepts, quantities, units and terminology for non-ionizing radiation protection", in *Health Physics*, 49, pp. 1329-1362; or in IRPA (1991), pp. 8-41.

—. 1990. "Interim guidelines on limits of exposure to 50/60 Hz electric and magnetic fields", in *Health Physics*, 58, pp. 113-122; or in IRPA (1991), pp. 83-94.

Kabrhel, I. 1985. *Answers to the CIGRE questionnaire "Survey of the situation concerning the effects of electric and magnetic fields"*, doc. 36-85 (SC) 28 IWD. Paris, CIGRE.

Kaune, W. T.; Forsythe, W. C. 1985. "Current densities measured in human models exposed to 60 Hz electric fields", in *Bioelectromagnetics*, 6, pp. 13-32.

—; —. 1988. "Current densities induced in swine and rat models by power-frequency electric fields", in *Bioelectromagnetics*, 9, pp. 1-24.

—; Phillips, R. D. 1980. "Comparison of the coupling of grounded humans, swine and rats to vertical 60 Hz electric fields", in *Bioelectromagnetics* 1, pp. 117-129.

Kavet, R. 1979. *Biological effects of high-voltage electric fields: An update*, Vol. 2: *Bibliography*. EPRI report EA 1123. Palo Alto, California, EPRI.

Knave, B., et al. 1979. "Long-term exposure to electric fields: A cross-sectional epidemiologic investigation of occupationally exposed workers in high-voltage substations", in *Scandinavian Journal of Work, Environment and Health*, 5, pp. 115-125.

Kornberg, H.; Sagan, L. 1979. *Biological effects of high-voltage fields: An update*. Vol. 1, EPRI report EA 1123. Palo Alto, California, EPRI.

Krause, N. 1986. "Exposure of people to static and time variable magnetic fields in technology, medicine, research and public life: Dosimetric aspects", in J. H. Bernhardt (ed.). *Biological effects of static and ELF-magnetic fields*, BGA-Schriftenreihe 3/86., Munich, MMV, pp 57-71.

Lövsund, P. 1981. *Biological effects of alternating current magnetic fields with special reference to the visual systems*. Linköping University. Linköping, Dissertation #47.

Lövsund, P., et al. 1982. "ELF magnetic fields in electrosteel and welding industries", in *Radio Science*, 17 (5S), pp. 35S-38S.

Lyskov, Y. I., et al. 1975. "The factors of electrical fields that have an influence on a human", in *Three Russian papers on EHV/UHV transmission line and substation design*. Boston, Uhl, Hall and Rich, Division of Charles T. Main, Inc.

Matthes, R.; Bernhardt, J. H. 1988. "Evaluation of the interference of electric and magnetic fields with the performance of unipolar cardiac pacemakers", in *Proceedings of the IVth European and the XIIIth Regional Congress of IRPA*. Seibersdorf, Österreichischer Verlag für Strahlenschutz, pp. 105-109.

Milham, S. 1982. "Mortality from leukemia in workers exposed to electrical and magnetic fields", in *New England Journal of Medicine*, 307, p. 249.

Miller, D. A. 1974. "Electric and magnetic fields produced by commercial power systems", in J. Llaurado et al. (eds.): *Biological and clinical effects of low-frequency magnetic and electric fields*. Springfield, III, C. C. Thomas, pp. 62-70.

NH & MRC (National Health and Medical Research Council). 1989. *Interim guidelines on limits of exposure to 50/60 Hz electric and magnetic fields*, Radiation Health Series No. 30, Australia, Canberra, Dec.

NRPB (National Radiological Protection Board, United Kingdom). 1989. *Guidance on standards: Guidance as to restrictions on exposure to time varying electromagnetic fields and the 1988 recommendations of the International Non-Ionizing Radiation Committee*, NRPB-GS11. Chilton, United Kingdom.

—. 1992. *Electromagnetic fields and the risk of cancer. Documents of the NRPB* (Chilton, United Kingdom), Vol. 3, No. 1.

Pilatowicz, A. 1985. *Answers to the CIGRE Questionnaire, "Survey of the situation concerning the effects of electric and magnetic fields"*, doc. 36-85 (SC) 09 IWD. Paris, CIGRE.

Polk, C.; Postow, E. 1986. *CRC Handbook of biological effects of electromagnetic fields*. Boca Raton, Florida, CRC Press.

Reilly, J. P.; Larkin, W.D. 1983. "Electrocutaneous stimulation with high-voltage capacitive discharges", in *IEEE Transactions on Biomedical Engineering*, BME-30, pp. 631-641.

Reilly, J. P., et al. 1981-85. *Human reactions to transient electric currents* (Laurel, Maryland, Johns Hopkins University, Applied Physics Laboratory): Annual Report CPE 8203, July 1982 (NTIS No. PB 83-204628); Annual Report CPE 8305, July 1983 (NTIS No. PB 84-112895); Annual Report CPE 8313, July 1984 (NTIS No. PB 84-231463); and J. P. Reilly and W. D. Larkin: *Human reactions to transient electric currents – Summary report*, PPSE T- 34, June 1985 (NTIS No. PB 86-117280). Available from the US National Technical Information Service, Springfield, Virginia.

Repacholi, M. H. 1985. "Standards on static and ELF electric and magnetic fields and their scientific basis", in M. Grandolfo et al., 1985.

—. 1992. "Guidelines and standards", in M. W. Greene (ed.): *Non-ionizing radiation*, Proceedings of the Second International Non-Ionizing Radiation Workshop, Vancouver, British Columbia. Canada, University of British Columbia Press, pp. 465-482.

Revnova, N. V., et al. 1968. "Effects of the high-intensity electric field of industrial frequency", in *Proceedings of the All-Union Symposium on the Hygiene of Labour and Biological Effects of Radiofrequency Electromagnetic Waves*.

Sander, R., et al. 1982. *Laboratory studies on animals and human beings exposed to 50 Hz electric and magnetic fields*, International Congress on Large High-Voltage Electric Systems, CIGRE paper 36-01. Paris, CIGRE.

Sazanova, T. E. 1967. *A physiological assessment of work conditions in 400-500 kV open switching yards*. Institute of Labour Protection of the All-Union Central of Trade Unions – Scientific Publication H 46, Profizdat. Translated by G. Knickerbocker in *Study in the USSR of medical effects of electric power systems*, IEEE Special Publication No. 10. Piscataway, New Jersey, IEEE Power Engineering Society.

Shah, K. R. 1979. *Review of state/federal environmental regulations pertaining to the electrical effects of overhead transmission lines: 1978*, Department of Energy Publications HCP/EV-1802. Washington, D.C, United States Department of Energy.

Sheppard, A. R.; Eisenbud, M. 1977. *Biological effects of electric and magnetic fields of extremely low frequency*. New York, New York University Press.

Sikov, M. R., et al. 1984. "Studies on prenatal and postnatal development in rats exposed to 60 Hz electric fields", in *Bioelectromagnetics*, 5, pp. 101-112.

Sikov, M. R., et al. 1985. *Biological studies of swine exposed to 60 Hz electric fields*, Vol. 4: *Growth, reproduction and development*, EPRI Report EA 4318. Palo Alto, California, EPRI.

Silny, J. 1986. "The influence thresholds of the time-varying magnetic fields in the human organism", in J. H. Bernhardt (ed.): *Biological effects of static and ELF-magnetic fields*, BGA-Schriftenreihe 3/86, Munich, MMV, pp. 105-112.

Silverman, C. 1990. "Epidemiological studies of cancer and electromagnetic fields", in O. P. Gandhi (ed.): *Biological effects and medical applications of electromagnetic energy*. New York, Prentice-Hall, pp. 415-436.

Singewald, M. L., et al. 1973. "Medical follow-up study of high-voltage linemen working in a.c. electric fields", in *IEEE Transactions on Power Apparatus and Systems*, PAS-92, pp. 1307-1309.

Spiegel, R. J. 1977. "Magnetic coupling to a prolate spheroid model of man", in *IEEE Transactions on Power Apparatus and Systems*, PAS-96(1), pp. 208-212.

—. 1981. "Numerical determination of induced currents in humans and baboons exposed to 60 Hz electric fields", in *IEEE Transactions on Electromagnetic Compatibility*, EMC-23, pp. 382-390.

Stollery, B. T. 1986. "Effects of 50 Hz electric currents on mood and verbal reasoning skills", in *British Journal of Medicine*, 43, p. 339.

Stuchly, M. A. 1986. "Exposure to static and time-varying magnetic fields in industry, medicine, research and public life: Dosimetric aspects", in J. H. Bernhardt (ed.): *Biological effects of static and ELF-magnetic fields*, BGA-Schriftenreihe 3/86. Munich, MMV, pp. 39-56.

Tell, R. A. 1983. "Instrumentation for measurement of radiofrequency electromagnetic fields: Equipment, calibrations, and selected applications", in M. Grandolfo et al. (eds.): *Biological effects and dosimetry of non-ionizing radiation: Radiofrequency and microwave energies*. New York, London, Plenum Press, pp. 95-162.

Tenforde, T. S. 1985. "Mechanisms for biological effects of magnetic fields", in M. Grandolfo et al. 1985, pp. 71-92.

—. 1986. "Magnetic field applications in modern technology and medicine", in J. H. Bernhardt (ed.): *Biological effects of static and ELF-magnetic fields*, BGA-Schriftenreihe 3/86. Munich, MMV, pp. 23-35.

—; Budinger, T. F. 1986. "Biological and physical aspects of NMR imaging and *in vivo* spectroscopy", in S. R. Thomas; R. L. Dickson (eds.): *NMR in medicine. The instrumentation and clinical applications*. New York, American Institute of Physics, pp. 493-548.

—; Kaune, W. T. 1987. "Interaction of extremely low frequency electric and magnetic fields with humans", in *Health Physics*, 53, pp. 585-606.

UNEP/WHO/IRPA (United Nations Environment Programme/World Health Organization/ International Radiation Protection Association). 1984. *Environmental Health Criteria 35, Extremely low frequency (ELF) fields*. Geneva, WHO.

—. 1987. *Environmental Health Criteria 69, Magnetic fields*. Geneva, WHO.

UNIPEDE (International Union of Producers and Distributors of Electrical Energy). 1985. *Rapport d'activité du Groupe d'études médicales*, Athens, 9-14 June 1985. Report 85.f.90.1.

USSR. 1970. *Rules and regulations on labour protection at 400, 500, and 750 kV AC substations and overhead lines of industrial frequency.* Moscow, USSR Ministry of Health. Translated by G. Knickerbocker, in *Study in the USSR of medical effects of electric power systems.* IEEE Special Publication No. 10. Piscataway, New Jersey, IEEE Power Engineering Society.

—. 1975. *Occupational Safety Standards System. Electrical fields of current industrial frequency of 400 kV and above. General safety requirements,* Standard No. 12.1.002-75. Moscow, National Standards Committee (in Russian).

—. 1985. *Maximum permissible levels of magnetic fields with the frequency 50 Hz,* doc. 3206-85. Moscow, USSR Ministry of Health (in Russian).

Wachtel, H. 1979. "Firing pattern changes and transmembrane currents produced by extremely low frequency fields in pacemaker neurons", in R. D. Phillips et al. (eds.): *Biological effects of extremely low frequency electromagnetic fields,* DOE Conference 781016, Washington, DC, US Department of Energy, pp. 132-146.

Wertheimer, N.; Leeper, E. 1979. "Electrical wiring configurations and childhood cancer", in *American Journal of Epidemiology,* 109, pp. 273-284.

WHO-EURO (World Health Organization Regional Office for Europe). 1989. *Non-ionizing radiation protection,* 2nd ed., edited by M. J. Suess and D. A. Benwell. Copenhagen.

Wilson, B. W., et al. 1981. "Chronic exposure to 60 Hz electric fields: Effects on pineal function in the rat", in *Bioelectromagnetics,* 2, pp. 371-380.

Yasui, M. 1985. *Answers to the CIGRE questionnaire "Survey of the situation concerning the effects of electric and magnetic fields",* doc. 36-85 (SC) 09 IWD. Paris, CIGRE.

Zaffanella, L. E. 1985. *Answers to the CIGRE questionnaire "Survey of the situation concerning the effects of electric and magnetic fields",* doc. 36-85 (SC) 21 IWD. Paris, CIGRE.

—; Deno, D. W. 1978. *Electrostatic and electromagnetic effects of ultra-high-voltage transmission lines.* Final report EPRI EL-802. Palo Alto, California, Electric Power Research Institute.